Mathematical
Surveys
and
Monographs

Volume 129

Steenrod Squares in Spectral Sequences

William M. Singer

American Mathematical Society

2000 *Mathematics Subject Classification*. Primary 16E40, 18G25, 18G30, 18G40, 55R20, 55R40, 55S10, 55T05, 55T15, 55T20.

For additional information and updates on this book, visit
www.ams.org/bookpages/surv-129

Library of Congress Cataloging-in-Publication Data

Singer, William M., 1942–
Steenrod squares in spectral sequences / William M. Singer.
p. cm. — (Mathematical surveys and monographs, ISSN 0076-5376 ; v. 129)
Includes bibliographical references and index.
ISBN 0-8218-4141-6 (alk. paper)
1. Spectral sequences (Mathematics) 2. Steenrod algebra. I. Singer, William M., 1942– II. Series: Mathematical surveys and monographs ; v. 129.

QA612.8 .S56 2006
515′.24—dc22 2006045953

∞ The paper used in this book is acid-free and falls within the guidelines
established to ensure permanence and durability.
Visit the AMS home page at http://www.ams.org/

10 9 8 7 6 5 4 3 2 1 11 10 09 08 07 06

Contents

Preface

In his thesis [**74**], J.-P. Serre initiated the use of spectral sequences in algebraic topology. Serre equipped his spectral sequence with a theory of products that greatly enhanced its effectiveness in computing the cohomology of a fiber space. Then it was natural to ask whether Steenrod's squaring operations could be introduced into spectral sequences as well. For example, in a collection of problems in algebraic topology [**46**] published in 1955, William Massey wrote: "Problem 6. Is it possible to introduce Steenrod squares and reduced p'th powers into the spectral sequence of a fibre mapping so that, for the term E_2, they behave according to the usual rules for squares and reduced p'th powers in a product space?"

The problem was first addressed by S. Araki [**6**] and R. Vazquez-Garcia [**87**], who both developed a theory of Steenrod operations in the Serre spectral sequence. L. Kristensen [**34**] also wrote down such a theory, and used it to calculate the cohomology rings of some two-stage Postnikov systems. Later, in the papers [**77, 78**], the present writer developed a theory of Steenrod operations for a class of first-quadrant cohomology spectral sequences. The theory applied to the Serre spectral sequence, but also to the change-of-rings spectral sequence for the cohomology of an extension of Hopf algebras, and to the Eilenberg-Moore spectral sequence for the cohomology of a classifying space.

The present work offers a more detailed exposition of the theory originally set out in [**77**] and [**78**], and generalizes its results.

To introduce the context for our theory, we recall the context in which Steenrod operations were originally defined, first by Steenrod [**83**], and then more generally by Dold [**16**]. One starts with a commutative, simplicial coalgebra R over the ground ring $\mathbb{Z}/2$. Associated to R is a chain complex CR, defined by $(CR)_n = R_n$, with differential the sum of the face operators of R. Steenrod and Dold showed how to use the "cup-i" products, $D_i : C(R \times R) \to CR \otimes CR$, together with the coproduct on R, to define the Steenrod squaring operations on the homology of the cochain complex dual to CR:

$$Sq^k : H^n(\mathrm{Hom}(CR, \mathbb{Z}/2)) \to H^{n+k}(\mathrm{Hom}(CR, \mathbb{Z}/2)). \tag{0.1}$$

To get a general theory of Steenrod operations in spectral sequences we generalize this construction in several directions. We begin by changing the ground ring from $\mathbb{Z}/2$ to any $\mathbb{Z}/2$-bialgebra Θ with commutative coproduct. This generalization is useful in dealing with Steenrod operations on the cohomology rings of Hopf algebras. Because we wish to discuss spectral sequences, we replace the simplicial coalgebra R by a bisimplicial object X over the category of Θ-coalgebras. This bisimplicial object determines a "total" chain complex CX of Θ-modules by the rule $(CX)_n = \bigoplus_{p+q=n}(X_{p,q})$. The differential is defined as a sum of the horizontal and vertical face operators in the usual way. We define Steenrod operations on the homology of the "dual" cochain complex, replacing the coefficient ring $\mathbb{Z}/2$ of (0.1) by an arbitrary Θ-algebra N:

$$Sq^k : H^n(\mathrm{Hom}_\Theta(CX, N)) \to H^{n+k}(\mathrm{Hom}_\Theta(CX, N)). \tag{0.2}$$

This construction is completely analogous to (0.1). But the bisimplicial object X also determines a double chain complex CX, with $(CX)_{p,q} = X_{p,q}$, and vertical and horizontal differentials defined as sums, respectively, of vertical and horizontal face operators. So a naturally defined first-quadrant spectral sequence ("the spectral sequence of a bisimplicial coalgebra") converges to the groups $H^*(\mathrm{Hom}_\Theta(CX, N))$ that appear in (0.2). We show how to put Steenrod operations into this spectral sequence. That is, for all $r \geq 2$ and all $p, q, k \geq 0$ we define Steenrod squares:

$$Sq^k : E_r^{p,q} \to E_r^{p,q+k} \qquad \text{if} \quad 0 \leq k \leq q, \tag{0.3a}$$

$$Sq^k : E_r^{p,q} \to E_t^{p+k-q,2q} \qquad \text{if} \quad q \leq k. \tag{0.3b}$$

Here t is an integer that must in general be chosen larger than r, in order that Sq^k be well-defined (the exact value of t is given in Theorem 2.15). However, if $r = 2$ then $t = 2$. All Steenrod squares defined on E_2 take values in E_2, and they carry a typical class $\alpha \in E_2^{p,q}$ to positions on the $P - Q$ plane as shown by the diagram:

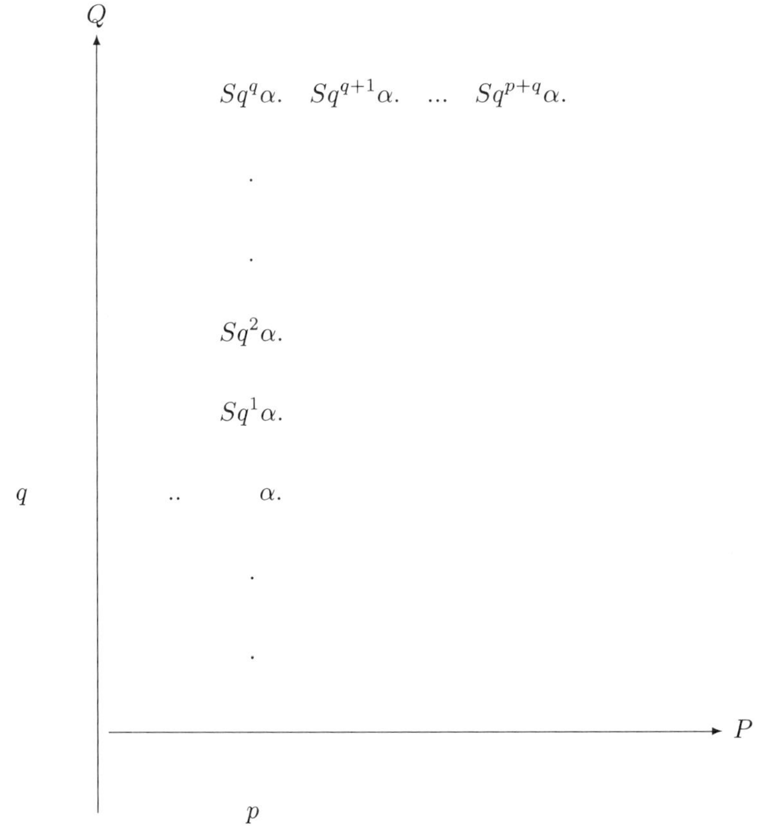

The squaring operations on E_r are determined by the operations on E_2, by passage to the subquotient (Theorem 2.16). For reasons made obvious by the diagram, we refer to the family (0.3a) as "vertical" operations, and the family (0.3b) as "diagonal". The Steenrod operations commute with the differentials of the spectral sequence (Theorem 2.17 and the diagram that follows it). The operations (0.3) are defined when $r = t = \infty$, and agree with those defined on the graded vector space associated to the filtration on the target groups $H^*(\mathrm{Hom}_\Theta(CX, N))$ of the spectral sequence (Theorem 2.22).

We apply the general theory in three cases. A. Dress observed [**17**] that the Serre spectral sequence can be obtained as the spectral sequence of a bisimplicial coalgebra. So our theory applies and we get Steenrod operations in the Serre spectral sequence, as in [**6, 34, 87**]. We sketch the application to the Serre spectral sequence at the end of Chapter 2, but do not develop it fully.

We work out our other examples in greater detail. Suppose Λ is a cocommutative Hopf algebra over $\mathbb{Z}/2$; suppose M is a commutative Λ-coalgebra and N a commutative Λ-algebra. In [**40**] Liulevicius showed how one can use cup-i products in a projective resolution of the Λ-module M to define Steenrod squares:

$$Sq^k : \mathrm{Ext}^p_\Lambda(M, N) \to \mathrm{Ext}^{p+k}_\Lambda(M, N). \tag{0.4}$$

These operations make $\mathrm{Ext}^*_\Lambda(M, N)$ a module over an algebra we call $\boldsymbol{\mathcal{H}}$. This is the algebra generated by the Steenrod squares and subject to the Adem relations, with the understanding that Sq^0 is not the unit of the algebra, but an independent generator. Now suppose given Γ, a normal sub-Hopf algebra of Λ, and let $\Omega = \Lambda//\Gamma$ be the quotient Hopf algebra. Then for any right Ω-module P and right Λ-modules Q, N we have the change-of-rings (Cartan-Eilenberg) spectral sequence :

$$E_2^{p,q} = \mathrm{Ext}^p_\Omega(P, \overline{\mathrm{Ext}}^\Gamma_{-q}(Q, N)) \Longrightarrow \mathrm{Ext}^{p+q}_\Lambda(P \otimes Q, N). \tag{0.5}$$

Here $\overline{\mathrm{Ext}}^\Gamma_*(Q, N)$ is the negatively graded cohomology of Γ, as explained in the text. If in addition P and Q are coalgebras over Ω and Λ respectively, and N is a commutative algebra over Λ, then the target of this spectral sequence supports Steenrod operations, as in (0.4). In this case we describe (0.5) as the spectral sequence of a bisimplicial coalgebra, so that we can apply our theory to give Steenrod squares in the spectral sequence. The work generalizes that of Uehara [**86**], who considered the case in which Γ is central in Λ and coefficient modules are $P = Q = N = \mathbb{Z}/2$. As in (0.3) we have both vertical and horizontal operations at the E_2 level:

$$Sq^k : \mathrm{Ext}^p_\Omega(P, \overline{\mathrm{Ext}}^\Gamma_{-q}(Q, N)) \to \mathrm{Ext}^p_\Omega(P, \overline{\mathrm{Ext}}^\Gamma_{-q-k}(Q, N)) \qquad (k \leq q) \tag{0.6a}$$

$$Sq^k : \mathrm{Ext}^p_\Omega(P, \overline{\mathrm{Ext}}^\Gamma_{-q}(Q, N)) \to \mathrm{Ext}^{p+k-q}_\Omega(P, \overline{\mathrm{Ext}}^\Gamma_{-2q}(Q, N)) \qquad (q \leq k). \tag{0.6b}$$

We give descriptions of both types of operation in terms of the actions of $\boldsymbol{\mathcal{H}}$ on the cohomologies of Γ and Ω. Thus, the vertical operations (0.6a) come about because $\overline{\mathrm{Ext}}^\Gamma_*(Q, N))$ supports an action of $\overline{\boldsymbol{\mathcal{H}}}$, the negative of $\boldsymbol{\mathcal{H}}$, that is "compatible" with the action of Ω, in a sense that we make precise in Chapters 3 and 5. We then use relative homological algebra in the category of $\overline{\boldsymbol{\mathcal{H}}} - \Omega$ modules to define these vertical operations. On the other hand we show that, after a shift in indexing, the diagonal operations (0.6b) are the same as those defined on the cohomology of the

Hopf algebra Ω using cup-i products on a projective resolution of the Ω-module P, as in (0.4).

Similar patterns appear in our study of the Eilenberg-Moore spectral sequence. Here we suppose given a simplicial group $\mathcal{G}$, a principal right $\mathcal{G}$-space $\mathcal{E}$ that is also a Kan complex, and a Kan complex $\mathcal{F}$ on which $\mathcal{G}$ acts from the left. The spectral sequence converges to the cohomology of the Borel construction:

$$E_2^{p,q} = \mathrm{Ext}^{p,q}_{H_*\mathcal{G}}(H_*\mathcal{E}, K_*\mathcal{F}) \Rightarrow H^{p+q}(\mathcal{E} \times_{\mathcal{G}} \mathcal{F}). \tag{0.7}$$

Here $K_*\mathcal{F}$ is the negatively graded cohomology of $\mathcal{F}$: $K_{-n}\mathcal{F} = H^n(\mathcal{F}, \mathbb{Z}/2)$. In Chapters 6 and 7 we obtain this spectral sequence as the spectral sequence of a bisimplicial coalgebra. Thus our general theory applies, and gives Steenrod squares in the spectral sequence. The vertical and horizontal operations on the E_2 term are:

$$Sq^k : \mathrm{Ext}^{p,q}_{H_*\mathcal{G}}(H_*\mathcal{E}, K_*\mathcal{F}) \to \mathrm{Ext}^{p,q+k}_{H_*\mathcal{G}}(H_*\mathcal{E}, K_*\mathcal{F}) \qquad (k \leq q) \tag{0.8a}$$

$$Sq^k : \mathrm{Ext}^{p,q}_{H_*\mathcal{G}}(H_*\mathcal{E}, K_*\mathcal{F}) \to \mathrm{Ext}^{p+k-q,2q}_{H_*\mathcal{G}}(H_*\mathcal{E}, K_*\mathcal{F}) \qquad (k \geq q). \tag{0.8b}$$

We give descriptions of these that are analogous to the descriptions we give for the operations on the change-of-rings spectral sequence. Thus, the vertical operations (0.8a) come about because both $K_*\mathcal{F}$ and $H_*\mathcal{E}$ support actions of $\overline{\boldsymbol{\mathcal{A}}}$, the negative of the Steenrod algebra $\boldsymbol{\mathcal{A}}$, that are compatible with the actions of $H_*\mathcal{G}$ in particular ways that we describe in Chapters 3 and 7. We then use relative homological algebra in the category of $\overline{\boldsymbol{\mathcal{A}}} - H_*\mathcal{G}$ modules to describe the operations (0.8a). On the other hand we show that, after a shift in indexing, the diagonal operations (0.8b) are the same as those defined on the cohomology of the Hopf algebra $H_*\mathcal{G}$ using cup-i products on a projective resolution of $H_*\mathcal{E}$, as in (0.4).

We discuss some applications that are already in the literature. These cite [**6, 77, 87**] or [**78**]. For example, Goerss [**25**] uses Steenrod squares in spectral sequences to discuss the André-Quillen cohomology of commutative algebras - a topic motivated by a desire to understand the E_2-term of the unstable Adams spectral sequence. We sketch this application at the end of Chapter 2. Palmieri [**64**] uses Steenrod operations in the change-of-rings spectral sequence in his recent work describing the cohomology of the Steenrod algebra "modulo nilpotence". We sketch this application at the end of Chapter 5. Mimura, Kameko, Kuribayashi, Mori, Nishimoto and Sambe [**31, 38, 53, 54, 55, 56, 57, 61, 62, 70, 71, 72**] have applied Steenrod operations in the Eilenberg-Moore spectral sequence to calculate the cohomology of the classifying spaces of the exceptional Lie groups. We cite their work at the end of Chapter 7.

We also discuss potential applications. There are several problems, or types of problem, to which the methods of this work can almost surely be applied to good effect. For example, at the end of Chapter 2 we discuss possible application to the Serre spectral sequence, in the case in which the cohomology of the fiber is not a simple coefficient system over the base. At the end of Chapter 5 we discuss possible applications of Steenrod squares in the change-of-rings spectral sequence to computing cohomologies of sub and quotient Hopf algebras of the Steenrod algebra, as well as to computing the cohomology rings of groups. Finally, we note that the operations (0.8) in the Eilenberg-Moore spectral sequence have so far been applied only in cases for which $\mathcal{E}$ is contractible and $\mathcal{F}$ is a point. It might be interesting, for example, to study cases in which $\mathcal{G}$ is a finite group, and $\mathcal{E}$ a product of spheres on which $\mathcal{G}$ acts freely.

Much of this work can be considered expository, but some of our results are new, and in particular go beyond [**77, 78**]. In [**78**] our treatment of the change-of-rings spectral sequence (0.5) relied on properties of extensions of Hopf algebras as they were developed in [**76**]. In particular we assumed that the Hopf algebras were graded-connected and that Γ had commutative multiplication. But in an effort to make the present work self-contained we have included Chapter 4, in which we develop from scratch the necessary properties of Hopf extensions. We are able to do this without assuming that the Hopf algebras are connected, and without assuming Γ commutative. Consequently, our results are more general than those in [**78**]. The greater generality is necessary, for example, for Palmieri's application [**64**].

Similarly, our discussion of the Eilenberg-Moore spectral sequence (0.7) is more detailed and general than the very brief treatment given in [**78**]. We begin in Chapter 6 by giving a self-contained construction of the spectral sequence. The original construction, in [**60**], used Serre's filtration of the chains of the total space $\mathcal{E}$, Serre's computation of the E_1-term of the resulting (Serre) spectral sequence, and the "comparison theorem for spectral sequences" that is developed in [**59**] and [**14**]. Our treatment does not use these results; but instead some properties of semi-simplicial fibrations from [**7**]. This approach is well suited to our purposes because it presents the Eilenberg-Moore spectral sequence directly as the spectral sequence of a bisimplicial coalgebra. So we can immediately apply the results of Chapter 2 to obtain Steenrod operations in the spectral sequence. These ideas were implicit in [**78**], but not worked out in detail. The present work goes beyond [**78**] also in allowing $\mathcal{F}$ to be any left $\mathcal{G}$-space, not necessarily a point. Our description of the Steenrod operations (0.8a) at the E_2 level, using relative homological algebra in the category of $\overline{\mathcal{A}} - H_*\mathcal{G}$ modules, also seems to be new.

In his papers [**61, 62**] Mori, influenced in part by [**77, 78**], has given a treatment of Steenrod operations in the Eilenberg-Moore spectral sequence at an arbitrary prime. Our present treatment has some features not found in [**61, 62**]. Our derivation of the spectral sequence is self-contained, and we get the theory of Steenrod operations within that setting. Our description of the operations at the E_2-level is independent of choices of resolutions.

Finally we use the present exposition to correct a mistake in a previous one. In [**77**] we asserted that the diagonal Steenrod squares (0.3b) are well defined as operations from E_r to E_r. This is not so if $r > 2$; at least, not in the theory we have developed. The mistake was pointed out by Sawka, in his treatment [**73**] of the odd primary version of [**77**]. In Chapter 2 of the present work we develop the general theory with great care, showing explicitly the verifications that the operations are well-defined, and giving equal attention to the related issue of their additivity.

All the results in this work have analogues at odd primes. We do not write these down explicitly. But, as we have just mentioned, Sawka's paper [**73**] develops the general theory of odd primary Steenrod operations in the spectral sequence of a bisimplicial coalgebra: it is the analogue of [**77**] and Chapter 2 of the present work. A reader interested particularly in change-of-rings or Eilenberg-Moore spectral sequences at odd primes should have no trouble combining Sawka's paper with the theorems in this book to get the results he needs. To facilitate such a process we write much of the present work over arbitrary characteristic, specializing to $p = 2$ only when the Steenrod operations are discussed. For example, Chapter 3 presents a theory of bialgebra actions on the cohomology of algebras that we need in several contexts. The characteristic is arbitrary and we pay all due attention to signs. The same is true of our treatment of extensions of Hopf algebras in Chapter 4, and of our derivation of the Eilenberg-Moore spectral sequence in Chapter 6.

This work treats only first-quadrant spectral sequences. Theories of Steenrod operations in second-quadrant spectral sequences have been developed by Smith [**80, 81**], Rector [**69**], and Dwyer [**18**]. In particular these theories apply to the second-quadrant Eilenberg-Moore spectral sequence that converges to the cohomology of a loop space; or, more generally, to the cohomology of an induced fibration.

The present work is organized in the following way. Chapter 1 is introductory. This chapter reviews some material concerning bialgebras and their modules, chain and cochain complexes, differential algebras, adjunction isomorphisms, cup-i products, Steenrod operations, relative homological algebra, and spectral sequences. Almost all of this is well-known; but we present the material in the forms in which we will be using it. One possibly novel feature is our notion of an "algebra with coproducts": a kind of generalized bialgebra. It seems to be the right description of the algebra $\mathcal{H}$ of Steenrod squares that operates on the cohomology rings of Hopf algebras. Further discussion of algebras with coproducts can be found in [**79**].

In Chapter 2 we develop the general theory of Steenrod operations in the spectral sequence of a bisimplicial coalgebra, and discuss some applications of this theory.

In Chapter 3 we give a set of axioms under which a bialgebra can operate on the cohomology of an algebra. We study some properties of this action; for example, its relationship to products and Steenrod squares, when these are defined on the cohomology of the algebra. We make several applications of this theory in later chapters. We use it to define the action of the base of an extension of Hopf algebras on the cohomology of the fiber, and to relate this action to the product and Steenrod squares on the cohomology of the fiber. The results play a role in the definitions of both vertical and diagonal operations (0.6a) and (0.6b) in Chapter 5. We use the theory of Chapter 3 also in Chapter 7 to define the vertical operations (0.8a) in the Eilenberg-Moore spectral sequence, and to relate them to the diagonal operations.

In Chapter 4 we develop some properties of extensions of Hopf algebras, and set up the change-of-rings spectral sequence, obtaining it as the spectral sequence of a bisimplicial coalgebra. Then in Chapter 5 we put products and Steenrod operations into the change-of-rings spectral sequence, and describe the operations at the E_2-level. We discuss some applications of these results.

In Chapter 6 we derive the Eilenberg-Moore spectral sequence "from scratch", obtaining it as the spectral sequence of a bisimplicial coalgebra. Then in Chapter 7 we define products and Steenrod operations in the spectral sequence, and describe the operations at the E_2-level. We mention some applications of these results.

The dependence of the various chapters upon one another is indicated by the diagram on the next page. Although the diagram shows Chapter 1 prerequisite for all other chapters, some sections of that chapter are needed only for the change-of-rings spectral sequence, and some are needed only for Eilenberg-Moore. In such cases we have so indicated at the beginning of the section.

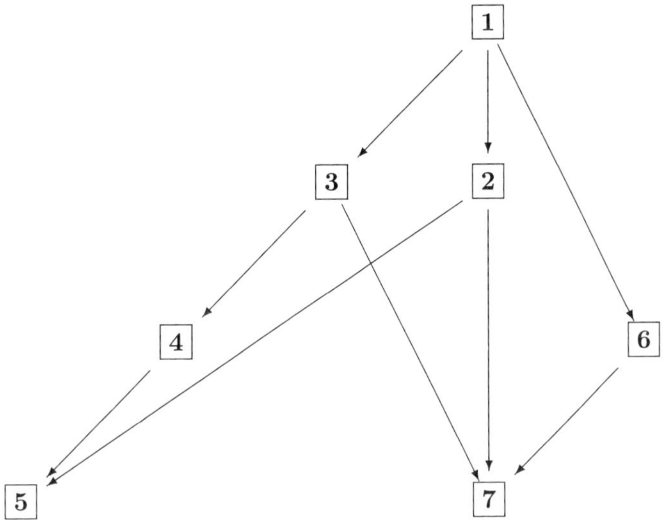

I would like to thank several persons for the assistance and encouragement they have given me while this book was being written.

Paul Taylor wrote "Commutative diagrams in TEX". Most of the diagrams in this book were created with the aid of this package.

I thank Peter Mix, science librarian at Fordham University, and Carol Hutchins, librarian of the Courant Institute of Mathematical Sciences, for their unfailing courtesy and helpfulness in locating materials I needed.

John Palmieri asked whether the results of [**78**] could be generalized to extensions of Hopf algebras with non-commutative kernel, thus providing me with one of the motivations for writing this book. William Dwyer, Philip Hirschhorn, and Daniel Kan gave valuable assistance with simplicial and bisimplicial theory. Dagmar Meyer read an early version of Chapter 2 and made many useful suggestions, all of which I have followed. I am grateful to Mamoru Mimura for several conversations and emails about Steenrod operations in the Eilenberg-Moore spectral sequence, and the role they have played in the computation of the cohomology rings of the classifying spaces of Lie groups.

Some forty years ago I began learning algebraic topology from John C. Moore, when he agreed to supervise my doctoral thesis. I have been learning from him ever since. His works on Hopf algebras, relative homological algebra, semi-simplicial fiber bundles, and the Eilenberg-Moore spectral sequence are major influences on this book.

I thank my wife, Eileen, for her patience and steady encouragement over the many months it took to write this book.

CHAPTER 1

Conventions

This chapter is a review of some basic material, and will serve to establish conventions that hold throughout the book. A reader experienced in algebraic topology should probably not read this chapter consecutively, but rather refer back to it as needed. Some sections pertain only to the change-of-rings spectral sequence, and some only to Eilenberg-Moore. In these cases we have so specified at the beginning of the section.

One possibly novel feature of this chapter is the notion of an "algebra with coproducts", discussed in Section 6. (See also the discussion in [**79**].) This seems to be the right way to describe the algebra $\boldsymbol{\mathcal{H}}$ that acts on the cohomology rings of Hopf algebras.

1. Vector Spaces

We fix a ground field l, and will consider $\mathbb{S}$-graded vector spaces

$$V = \{V_s \mid s \in \mathbb{S}\} \tag{1.1}$$

where $\mathbb{S}$ is an abelian group, and each V_s is an ungraded vector space over l. In our work, $\mathbb{S}$ will be either 0 or $\mathbb{Z}^n$, an n-fold product of copies of the integers. We will assume that in every case S comes equipped with a particular homomorphism of groups $\rho : \mathbb{S} \to \mathbb{Z}$. If $v \in V_s$ then the integer $\rho(s)$ will be called *total degree* of v, and we will write $|v| = \rho(s)$. We must keep track of total degree: we will use it to define the signs associated with certain homomorphisms of vector spaces (immediately below, for example, where we discuss switching factors in a tensor product). When we say that two vector spaces are graded on the same group $\mathbb{S}$, we mean that the homomorphisms $\rho : \mathbb{S} \to \mathbb{Z}$ are also the same.

We will sometimes write the vector space V of (1.1) as V_*; say, if there is reason to emphasize the grading. Often a vector space V will be graded on a cross product $\mathbb{S} \times \mathbb{T}$ of two abelian groups. We may write $V_{*,*}$ for such a vector space; or just V_* if only one of the gradings is important in context. The following definition is standard:

Definition 1.1. The graded vector space (1.1) is called "locally finite" if V_s is finite-dimensional for each $s \in \mathbb{S}$.

Given a family $\{V^{(i)} \mid i \in \mathcal{I}\}$ of $\mathbb{S}$-graded vector spaces, indexed by a set $\mathcal{I}$, their sum and product, respectively, are the $\mathbb{S}$-graded vector spaces defined by:

$$\left(\bigoplus\nolimits_{i\in\mathcal{I}} V^{(i)}\right)_s = \bigoplus_{i\in\mathcal{I}} V_s^{(i)}$$

$$\left(\prod_{i\in\mathcal{I}} V^{(i)}\right)_s = \prod_{i\in\mathcal{I}} V_s^{(i)}.$$

DEFINITION 1.2. Let V be an $\mathbb{S}$-graded vector space. By its "negative", $\overline{V}$, we mean the $\mathbb{S}$-graded vector space defined by:

$$\overline{V}^s = V_{-s}, \qquad \forall s \in S.$$

We adopt the convention that the homomorphism $\rho : \mathbb{S} \to \mathbb{Z}$ that defines total degree in $\overline{V}$ is the same as the homomorphism ρ that defines total degree in V.

In what follows, the unadorned tensor product is always over ground field l.

DEFINITION 1.3. Let V, W be vector spaces graded on abelian groups $\mathbb{S}, \mathbb{T}$, respectively. By the uncontracted tensor product $V\widehat{\otimes}W$ we mean the vector space graded on $\mathbb{S}\times\mathbb{T}$ defined by

$$(V\widehat{\otimes}W)_{s,t} = V_s \otimes W_t$$

for all $s \in \mathbb{S}$, $t \in \mathbb{T}$. Total degree is defined by the formula $|v\otimes w| = |v| + |w|$.

DEFINITION 1.4. Suppose V, W are vector spaces graded on the same group $\mathbb{S}$. By the (ordinary, contracted) tensor product $V \otimes W$ we mean the $\mathbb{S}$-graded vector space defined by

$$(V\otimes W)_s = \bigoplus_{\{i.j\in\mathbb{S}|i+j=s\}} (V_i \otimes W_j)$$

for each $s \in \mathbb{S}$. In particular, total degree in the tensor product is determined by $|v\otimes w| = |v| + |w|$.

Given vector spaces $V_1, \ldots, V_n$ and a permutation π of the symbols $\{1, 2, \ldots, n\}$, we will write

$$(\pi(1), \pi(2), \ldots, \pi(n)) : V_1 \otimes V_2 \otimes \cdots \otimes V_n \to V_{\pi(1)} \otimes V_{\pi(2)} \otimes \cdots \otimes V_{\pi(n)} \tag{1.2}$$

for the homomorphism that permutes the factors in a tensor product. In the graded case in which we are working, switching factors involves multiplication by ± 1. Thus $(2,1) : V\otimes W \to W \otimes V$ is given by $v\otimes w \to (-1)^{|v||w|} w \otimes v$. More general permutations are written as products of transpositions and the associated signs computed accordingly.

DEFINITION 1.5. Supposing V, W vector spaces graded on groups $\mathbb{S}, \mathbb{T}$, respectively, then by a homomorphism $f : V \to W$ we mean a mapping of sets $\phi : \mathbb{S} \to \mathbb{T}$, together with a family of (ungraded) homomorphisms of vector spaces $\{f_k : V_k \to W_{\phi(k)} \,|\, k \in \mathbb{S}\}$. The homomorphism f is called homogeneous if V and W are graded on the same group $\mathbb{S}$, and if ϕ has the form $\phi(k) = k+s$ for some fixed $s \in \mathbb{S}$. Then s is called the degree of the homomorphism. In this case our definitions imply that f raises total degree by $\rho(s)$. When we speak of the category of $\mathbb{S}$-graded vector spaces, we mean that the morphisms are the homogeneous homomorphisms. A homomorphism that is homogeneous of degree 0 is called degree-preserving.

DEFINITION 1.6. Supposing V, W vector spaces graded on groups $\mathbb{S}, \mathbb{T}$, respectively, a homomorphism $f : V \to W$ is called parity-preserving if:

1. either $|f(v)| \equiv |v| \pmod 2$ for all $v \in V$, or:
2. the characteristic of the ground field is 2.

In what follows, unadorned "Hom" is always taken over the ground field l.

Suppose V, W are vector spaces graded on the same group $\mathbb{S}$. By $\mathrm{Hom}(V,W)$ and $\overline{\mathrm{Hom}}(V,W)$ we mean the $\mathbb{S}$-graded vector spaces defined by

$$\mathrm{Hom}^s(V,W) = \prod_{k\in\mathbb{S}} \mathrm{Hom}(V_{k+s}, W_k) \tag{1.3a}$$

$$\overline{\mathrm{Hom}}_s(V,W) = \prod_{k\in\mathbb{S}} \mathrm{Hom}(V_k, W_{k+s}) \tag{1.3b}$$

for each $s \in \mathbb{S}$. $\overline{\mathrm{Hom}}(V,W)$ is the negative of the vector space $\mathrm{Hom}(V,W)$, in the sense of Definition 1.2. The elements of $\mathrm{Hom}^{-s}(V,W)$, and of $\overline{\mathrm{Hom}}_s(V,W)$, are the homogeneous homomorphisms of degree s in the sense of Definition 1.5.

If V is an $\mathbb{S}$-graded vector space then $\mathrm{Hom}(V,\_\_)$ is a covariant functor from the category of $\mathbb{S}$-graded vector spaces and homogeneous maps to itself. If W is an $\mathbb{S}$-graded vector space then $\mathrm{Hom}(\_\_,W)$ is a contravariant functor from the category of $\mathbb{S}$-graded vector spaces and homogeneous maps to itself. As $\mathrm{Hom}(\_\_,\_\_)$ is a bifunctor, so is $\overline{\mathrm{Hom}}(\_\_,\_\_)$.

The convention (1.3a) is standard in algebraic topology. But the alternative grading (1.3b) is necessary for the description of adjunction isomorphisms in categories of graded vector spaces, a topic that we now review.

Supposing U, V, W vector spaces all graded on $\mathbb{S}$, we are going to define a pair of vector space isomorphisms, inverse to each other:

$$\mathrm{Hom}(U\otimes V, W) \underset{\zeta}{\overset{\phi}{\rightleftarrows}} \mathrm{Hom}(U, \overline{\mathrm{Hom}}(V,W)). \tag{1.4}$$

First, given $f \in \mathrm{Hom}^s(U\otimes V, W)$ we define $\phi f \in \mathrm{Hom}^s(U, \overline{\mathrm{Hom}}(V,W))$. We must specify the components $(\phi f)_k : U_{k+s} \to \overline{\mathrm{Hom}}_k(V,W)$ for all $k \in \mathbb{S}$. We do this, in turn, by defining for each $k \in \mathbb{S}$, all $u \in U_{k+s}$ and each $j \in \mathbb{S}$ the component of $(\phi f)_k(u)$ in $\mathrm{Hom}(V_j, W_{j+k})$. The definition is

$$[(\phi f)_k(u)](v) = f(u\otimes v) \tag{1.5}$$

for all $v \in V_j$. Of course the form of (1.5) is well known. We repeat it here in the case of graded vector spaces in order to verify that, with gradings defined by (1.3a) and (1.3b), ϕ is degree-preserving. We next define the inverse mapping ζ. For each $s \in \mathbb{S}$ we need to describe $\zeta : \mathrm{Hom}^s(U, \overline{\mathrm{Hom}}(V,W)) \to \mathrm{Hom}^s(U\otimes V, W)$. So for each $g \in \mathrm{Hom}^s(U, \overline{\mathrm{Hom}}(V,W))$ and each $k \in \mathbb{S}$ we must define the component $(\zeta g)_k \in \mathrm{Hom}((U\otimes V)_{k+s}, W_k)$. We do this by setting

$$(\zeta g)_k(u\otimes v) = [g(u)](v) \tag{1.6}$$

for all $u \in U_i$, $v \in V_j$, $i+j = k+s$. We will use (1.5) and (1.6) in Chapter 4.

The isomorphisms (1.5), (1.6) are natural with respect to homogeneous homomorphisms of the variables U, V, W in the obvious sense.

Supposing V, V', W, W' vector spaces graded on $\mathbb{S}$, we define degree-preserving homomorphisms of vector spaces

$$\nu : \mathrm{Hom}(V,W)\otimes \mathrm{Hom}(V',W') \to \mathrm{Hom}(V\otimes V', W\otimes W') \tag{1.7a}$$

$$\overline{\nu} : \overline{\mathrm{Hom}}(V,W)\otimes \overline{\mathrm{Hom}}(V',W') \to \overline{\mathrm{Hom}}(V\otimes V', W\otimes W') \tag{1.7b}$$

by writing

$$(\nu(f \otimes f'))(v \otimes v') = (-1)^{|f'||v|} f(v) \otimes f'(v'), \tag{1.8}$$

and similarly for $\overline{\nu}$. The sign, like the others in this book, has been determined by the rule that, whenever one interchanges two adjacent symbols, one multiplies by -1 raised to a power that is the product of the total degrees of the symbols.

REMARK 1.7. Since homogeneous components of graded homomorphism groups are defined in (1.3) as products, not sums, equations like (1.8) must be interpreted with care. One cannot say that (1.8) defines ν on a spanning set, and that one extends by linearity! Rather, given $f \in \mathrm{Hom}^k(V, W)$, $f' \in \mathrm{Hom}^{k'}(V', W')$, with $k + k' = s$ in $\mathbb{S}$, then f is a family of homomorphisms $f_i : V_{k+i} \to W_i$, one for each $i \in \mathbb{S}$; and f' is a family $f'_{i'} : V'_{k'+i'} \to W'_{i'}$, one for each $i' \in \mathbb{S}$. Then $\nu(f \otimes f')$ is being defined as a family of homomorphisms $(\nu(f \otimes f'))_p : (V \otimes V')_{p+s} \to (W \otimes W')_p$, one for each $p \in \mathbb{S}$. Equation (1.8) is a shorthand for: $(\nu(f \otimes f'))_p(v \otimes v') = (-1)^{|f'||v|} f_i(v) \otimes f'_{i'}(v')$ for all $v \in V_{k+i}$, $v' \in V'_{k'+i'}$, and for all i, i' for which $i + i' = p$. In describing homomorphisms into vector spaces of the form $\mathrm{Hom}(\_,\_)$ we will often employ this kind of shorthand.

DEFINITION 1.8. Let V be a vector space graded on $\mathbb{S} = \mathbb{Z}^n$, for some $n \geq 0$. We say V is bounded below if there exist integers $k_1, \cdots, k_n$ such that $V_{j_1,\cdots,j_n} = 0$ if $j_i < k_i$ for some i. The definition of an $\mathbb{S}$-graded vector space being bounded above is analogous.

REMARK 1.9. Suppose V and W are vector spaces graded on the same group $\mathbb{S}$. Suppose that V is locally finite and bounded below, and that W is bounded above. Then there is an obvious, natural, degree-preserving isomorphism of $\mathbb{S}$-graded vector spaces:

$$\mathrm{Hom}(V, W) \; = \; \mathrm{Hom}(V, l) \otimes \overline{W}. \tag{1.9}$$

2. Algebras, Coalgebras, and Modules

Given an abelian group $\mathbb{S}$, an $\mathbb{S}$-graded algebra is an $\mathbb{S}$-graded vector space Θ with associative multiplication $\mu : \Theta \otimes \Theta \to \Theta$ and two-sided unit $\eta : l \to \Theta$. Here it is understood that l is the $\mathbb{S}$-graded vector space concentrated in degree $0 \in \mathbb{S}$, and that the structure maps μ and η are degree-preserving. Similarly, $\mathbb{S}$-graded coalgebras Π have degree-preserving coproduct $\psi : \Pi \to \Pi \otimes \Pi$ and counit $\epsilon : \Pi \to l$. Units, multiplications, counits, comultiplications will always be designated by the Greek letters $\eta, \mu, \epsilon, \psi$, respectively; or, for example, as ψ_Π if more than one coalgebra is involved.

When we speak of commutativity of an algebra, or of a coalgebra, we mean in the graded sense. *We assume all comultiplications commutative,* and we have tried to point out in the course of our proofs where this assumption is necessary.

If Θ is an $\mathbb{S}$-graded algebra, then the vector space $\overline{\Theta}$, as given in Definition 1.2, inherits a unit and product from Θ in the obvious way. We refer to $\overline{\Theta}$ so equipped as the negative of the algebra Θ.

DEFINITION 1.10. The tensor product of the $\mathbb{S}$-graded algebras Θ and Θ' is an $\mathbb{S}$-graded algebra on the underlying vector space $\Theta \otimes \Theta'$ with product defined by the composition:

$$\Theta \otimes \Theta' \otimes \Theta \otimes \Theta' \xrightarrow{(1,3,2,4)} \Theta \otimes \Theta \otimes \Theta' \otimes \Theta' \xrightarrow{\mu_\Theta \otimes \mu_{\Theta'}} \Theta \otimes \Theta'.$$

We have repeated this well-known definition only to remind the reader of the sign introduced by the permutation of factors. Similarly, the tensor product of two $\mathbb{S}$-graded coalgebras is an $\mathbb{S}$-graded coalgebra.

If Π is a coalgebra and Θ an algebra, graded on the same group $\mathbb{S}$, the set $\mathrm{Hom}^0(\Pi, \Theta)$ of degree-preserving, l-linear maps $f : \Pi \to \Theta$ is a semigroup under convolution product, defined by

$$f * g = \mu(f \otimes g)\psi. \tag{1.10}$$

The identity element of this semigroup is the composition

$$\Pi \xrightarrow{\epsilon} l \xrightarrow{\eta} \Theta. \tag{1.11}$$

In what follows the reader should distinguish carefully between composition of maps and convolution of maps. The former will always be denoted by simple juxtaposition, and the latter with an asterisk as above. The symbol f^{-1} will be reserved for the two-sided inverse of a mapping under convolution product, if such inverse exists.

We review the definition of the "Verschiebung", a self-mapping that can be defined on any commutative coalgebra if the ground field is $l = \mathbb{Z}/2$. In fact, if Π is any $\mathbb{S}$-graded vector space over $\mathbb{Z}/2$ write $T : \Pi \otimes \Pi \to \Pi \otimes \Pi$ for the homomorphism that switches the order of the factors in the tensor product, and write $(\Pi \otimes \Pi)^T$ for the vector subspace of elements invariant under the action of T. We write $1 : \Pi \otimes \Pi \to \Pi \otimes \Pi$ for the identity mapping. Then the sum $1 + T$ can be regarded as a homomorphism $1 + T : \Pi \otimes \Pi \to (\Pi \otimes \Pi)^T$. Suppose $\{\pi_i \mid i \in I\}$ a basis for Π, where I is some indexing set. Then it is clear that each element ρ of $(\Pi \otimes \Pi)^T$ can be written uniquely as a $\mathbb{Z}/2$-linear combination:

$$\rho = \sum_{(i \neq i')} \lambda_{i,i'}(\pi_i \otimes \pi_{i'} + \pi_{i'} \otimes \pi_i) + \sum_j \kappa_j(\pi_j \otimes \pi_j). \tag{1.12}$$

Here the first sum is over all unordered pairs (i, i') of indices in I, the second is over all indices $j \in I$, and only finitely many of the coefficients $\lambda_{i,i'}$, κ_j are non-zero. Thus, each choice of basis for Π determines a direct sum splitting

$$(\Pi \otimes \Pi)^T = \mathrm{im}(1 + T) \oplus \Pi, \tag{1.13}$$

where the associated injection $\Pi \to (\Pi \otimes \Pi)^T$ is defined on the basis by $\pi_j \to \pi_j \otimes \pi_j$ for each $j \in I$.

LEMMA 1.11. *Let Π be a graded vector space over $\mathbb{Z}/2$. For each $\rho \in (\Pi \otimes \Pi)^T$ there exists a unique $\rho' \in \Pi$ having the property:*

$$\rho - \rho' \otimes \rho' \in \mathrm{im}(1 + T). \tag{1.14}$$

PROOF. To prove existence let $\rho \in (\Pi \otimes \Pi)^T$ be given, and choose a basis $\{\pi_i \mid i \in I\}$ for Π. Expand ρ as in (1.12) and set $\rho' = \sum_j \kappa_j \pi_j$. Since we are working over $\mathbb{Z}/2$ we have

$$\rho' \otimes \rho' = \sum_{(j \neq j')} \kappa_j \kappa_{j'}(\pi_j \otimes \pi_{j'} + \pi_{j'} \otimes \pi_j) + \sum_j \kappa_j(\pi_j \otimes \pi_j), \tag{1.15}$$

where the first sum is over all unordered pairs (j, j') of indices in I. So it is clear from (1.12) that (1.14) is satisfied. To prove uniqueness, suppose that in addition to (1.14) there is a vector $\rho'' \in \Pi$ for which

$$\rho - \rho'' \otimes \rho'' \in \operatorname{im}(1+T). \tag{1.16}$$

We suppose $\rho' \neq \rho''$ and will obtain a contradiction. Consider the case in which neither ρ' nor ρ'' is zero. Then we can extend the pair $\{\rho', \rho''\}$ to a basis for Π and construct the corresponding splitting (1.13). Then (1.14) says that $\rho' \otimes \rho'$ is the component of ρ that lies in the summand Π of (1.13), whereas (1.16) says that $\rho'' \otimes \rho''$ is that component. Since $\rho' \neq \rho''$ we have our contradiction. The case in which $\rho' = 0$ and $\rho'' \neq 0$ leads similarly to a contradiction, and our proof of the uniqueness of ρ' satisfying (1.14) is complete. □

DEFINITION 1.12. Suppose Π a commutative $\mathbb{S}$-graded coalgebra and the ground field is $\mathbb{Z}/2$. Given $\pi \in \Pi$, we define $V(\pi) \in \Pi$, to be the unique element for which

$$\psi(\pi) - V(\pi) \otimes V(\pi) \in \operatorname{im}(1+T).$$

Clearly $V : \Pi \to \Pi$ is a degree-halving homomorphism; this is the "Verschiebung". So for each $\pi \in \Pi$, we have an expansion the form:

$$\psi(\pi) = \sum_i (\pi_i \otimes \pi_i' + \pi_i' \otimes \pi_i) + V(\pi) \otimes V(\pi). \tag{1.17}$$

It is easy to check that

$$\psi V = (V \otimes V)\psi \tag{1.18}$$

as homomorphisms $\Pi \to \Pi \otimes \Pi$. If Υ and Π are both $\mathbb{S}$-graded coalgebras over $\mathbb{Z}/2$ then in the coalgebra $\Upsilon \otimes \Pi$ one has

$$V(\upsilon \otimes \pi) = V\upsilon \otimes V\pi \tag{1.19}$$

for all $\upsilon \in \Upsilon$, $\pi \in \Pi$.

We return to the case of general ground field.

DEFINITION 1.13. If Θ is an $\mathbb{S}$-graded algebra then by a right Θ-module we mean an $\mathbb{S}$-graded vector space M with degree-preserving action $\sigma : M \otimes \Theta \to M$ satisfying unitary and associative laws. Left modules are defined similarly.

Definitions and theorems in the remainder of this chapter will be given with the understanding that "module" means "right module"; the analogous statements for left modules will often be implicit (see however (1.22). We reserve the Greek letter σ for actions of an algebra Θ on a module M; or will write $\sigma_M : M \otimes \Theta \to M$ when more than one module is involved.

DEFINITION 1.14. Suppose Θ and Θ' are algebras graded on $\mathbb{S}$. By the tensor product of the Θ-module M with the Θ'-module M' we mean the $\Theta \otimes \Theta'$-module $M \otimes M'$ with action given by the composition:

$$M \otimes M' \otimes \Theta \otimes \Theta' \xrightarrow{(1,3,2,4)} M \otimes \Theta \otimes M' \otimes \Theta' \xrightarrow{\sigma_M \otimes \sigma_{M'}} M \otimes M'. \tag{1.20}$$

Like Definition 1.10 this is standard; we repeat it only to emphasize that the permutation of factors introduces a sign which must be remembered. It plays a role, for example, in the proof of Proposition 1.18.

DEFINITION 1.15. If Θ is an algebra and M a right Θ-module, we write $\mathcal{F}^{\Theta}(M)$ for the free Θ-module generated by M. The underlying vector space is $\mathcal{F}^{\Theta}(M) = M \otimes \Theta$, and the Θ-action is given by $(m \otimes \theta)\theta' = m \otimes \theta\theta'$.

Of course the definition of $\mathcal{F}^{\Theta}(M)$ as a Θ-module would make sense even if M were only a vector space over the ground field. But we will be applying this definition only to those M that are already Θ-modules.

DEFINITION 1.16. Let Θ be an $\mathbb{S}$-graded algebra and M, N a pair of right Θ-modules, and P, Q a pair of left Θ-modules. By a homomorphism of right Θ-modules we mean a homogeneous homomorphism $f : M \to N$ of graded vector spaces satisfying

$$f(m\theta) = f(m)\theta \tag{1.21}$$

for all $m \in M$, $\theta \in \Theta$. By a homomorphism of left Θ-modules we mean a homogeneous homomorphism $f : P \to Q$ of graded vector spaces satisfying

$$f(\theta p) = (-1)^{|\theta||f|}\theta f(p) \tag{1.22}$$

for all $p \in P$, $\theta \in \Theta$.

The sign in (1.22) will not play a big role in this book, and may seem unconventional; but it is necessary. See Remarks 1.19 and 3.5.

DEFINITION 1.17. Let Θ be an $\mathbb{S}$-graded algebra and M, N a pair of right Θ-modules. By $\mathrm{Hom}_{\Theta}(M, N)$ we mean the sub $\mathbb{S}$-graded vector space of $\mathrm{Hom}(M, N)$ consisting of those elements that are homomorphisms of Θ-modules. We define $\overline{\mathrm{Hom}}^{\Theta}(M, N)$ to be the sub $\mathbb{S}$-graded vector space of $\overline{\mathrm{Hom}}(M, N)$ consisting of those elements that are homomorphisms of Θ-modules.

PROPOSITION 1.18. *Suppose Θ and Θ' are algebras; suppose M and N are right modules over Θ, and that M' and N' are right modules over Θ'. Then the mappings ν of* (1.7) *restrict to:*

$$\nu : \mathrm{Hom}_{\Theta}(M, N) \otimes \mathrm{Hom}_{\Theta'}(M', N') \quad \to \quad \mathrm{Hom}_{\Theta\otimes\Theta'}(M \otimes M', N \otimes N') \tag{1.23a}$$

$$\overline{\nu} : \overline{\mathrm{Hom}}^{\Theta}(M, N) \otimes \overline{\mathrm{Hom}}^{\Theta'}(M', N') \quad \to \quad \overline{\mathrm{Hom}}^{\Theta\otimes\Theta'}(M \otimes M', N \otimes N'). \tag{1.23b}$$

This is an easy computation requiring only that one take note of the signs explicit in (1.8), implicit in (1.20), and our assumption that all actions are degree-preserving. We will use the pairings (1.23) in Chapters 2 and 3.

REMARK 1.19. Proposition 1.18 has an obvious analogue for left Θ-modules, which requires the sign in (1.22).

REMARK 1.20. Suppose Θ an augmented algebra; i.e., Θ is equipped with an algebra homomorphism $\epsilon : \Theta \to l$ which is left inverse to the unit $\eta : l \to \Theta$. Then one can define the notion of a "trivial" Θ-module: this is a Θ-module for which the action of Θ factors through the augmentation. If M is an arbitrary right Θ-module that is both locally finite and bounded below, and if N is a trivial right Θ-module that is bounded above, then the isomorphism (1.9) restricts to a degree-preserving isomorphism of $\mathbb{S}$-graded vector spaces:

$$\mathrm{Hom}_{\Theta}(M, N) \; = \; \mathrm{Hom}_{\Theta}(M, l) \otimes \overline{N}.$$

This isomorphism is natural with respect to homogeneous homomorphisms of the Θ-module M.

3. Dual Modules

If Θ is an algebra and M is a right Θ-module, the dual vector space $\mathrm{Hom}(M, l)$ becomes a left $\overline{\Theta}$-module under the action defined by

$$(\overline{\theta} \cdot f)(m) = (-1)^{|f(m)||\theta|} f(m\theta) \tag{1.24}$$

for all $f \in \mathrm{Hom}(M, l)$, $\overline{\theta} \in \overline{\Theta}$, $m \in M$. Here, for $\theta \in \Theta$, we are writing $\overline{\theta}$ for the corresponding element of $\overline{\Theta}$. Notice that our newly defined action $\overline{\Theta} \otimes \mathrm{Hom}(M, l) \to \mathrm{Hom}(M, l)$ is degree-preserving, as required by our definition of a module over an algebra; that is why we must have $\overline{\Theta}$ acting, rather than Θ.

PROPOSITION 1.21. *Let Θ and Θ' be $\mathbb{S}$-graded algebras, and suppose given right modules M and M' over Θ and Θ' respectively. Then the pairing*

$$\nu : \mathrm{Hom}(M, l) \otimes \mathrm{Hom}(M', l) \to \mathrm{Hom}(M \otimes M', l)$$

as in (1.7a), *is a homomorphism of left $\overline{\Theta} \otimes \overline{\Theta'}$-modules.*

This is easy to check.

Similarly if P is a left Θ-module, the dual vector space $\overline{\mathrm{Hom}}(P, l)$ becomes a right Θ-module under the action of Θ defined by

$$(f \cdot \theta)(p) = f(\theta p) \tag{1.25}$$

for all $f \in \overline{\mathrm{Hom}}(P, l)$, all $\theta \in \Theta$, and $p \in P$.

PROPOSITION 1.22. *Let Θ and Θ' be $\mathbb{S}$-graded algebras, and suppose given left modules P and P' over Θ and Θ' respectively. Then the pairing*

$$\overline{\nu} : \overline{\mathrm{Hom}}(P, l) \otimes \overline{\mathrm{Hom}}(P', l) \to \overline{\mathrm{Hom}}(P \otimes P', l)$$

as in (1.7b), *is a homomorphism of right $\Theta \otimes \Theta'$-modules.*

4. Bialgebras and Hopf Algebras

A bialgebra Θ graded on the abelian group $\mathbb{S}$ is both an algebra and a coalgebra over $\mathbb{S}$: it is required that both the counit $\epsilon : \Theta \to l$ and coproduct $\psi : \Theta \to \Theta \otimes \Theta$ are morphisms of algebras; and it is required that both the unit $\eta : l \to \Theta$ and product $\mu : \Theta \otimes \Theta \to \Theta$ are morphisms of coalgebras.

The identity element $\eta\epsilon$ of the semigroup $\mathrm{Hom}^0(\Theta, \Theta)$ is as in (1.11) and must be distinguished from the identity mapping $\Theta : \Theta \to \Theta$ from Θ to itself. If the latter has a two-sided inverse in the semigroup then Θ is by definition a Hopf algebra. This two-sided inverse is traditionally called the antipode, and we will write it $\chi = \Theta^{-1}$ in $\mathrm{Hom}^0(\Theta, \Theta)$. So by definition the following diagram commutes:

$$(1.26)\qquad \begin{array}{ccccc} \Theta & \xrightarrow{\psi} & \Theta\otimes\Theta & \xrightarrow{\chi\otimes\Theta} & \Theta\otimes\Theta \\ \downarrow{\scriptstyle\psi} & \searrow{\scriptstyle\eta\epsilon} & & & \\ \Theta\otimes\Theta & & & & \downarrow{\scriptstyle\mu} \\ \downarrow{\scriptstyle\Theta\otimes\chi} & & & & \\ \Theta\otimes\Theta & & \xrightarrow[\mu]{} & & \Theta. \end{array}$$

The antipode is an antiautomorphism of both algebras and coalgebras, as shown in ([**58**, Proposition 1.5.10]). In particular the antipode respects both unit and counit in the sense that $\chi\eta = \eta; \epsilon\chi = \epsilon$.

If Θ is a bialgebra over $\mathbb{Z}/2$ then the Verschiebung satisfies

$$V\mu = \mu(V \otimes V) \tag{1.27}$$

as homomorphisms $\Theta \otimes \Theta \to \Theta$. If Θ is a Hopf algebra over $\mathbb{Z}/2$ then:

$$V\chi = \chi V \tag{1.28}$$

as homomorphisms $\Theta \to \Theta$.

We return to the case of arbitrary ground field.

If Θ is an $\mathbb{S}$-graded bialgebra or a Hopf algebra, then the $\mathbb{S}$-graded vector space $\overline{\Theta}$ inherits from Θ the structure of a bialgebra (respectively, Hopf algebra), in the obvious way. We refer to $\overline{\Theta}$ so equipped as the negative of the bialgebra Θ (respectively, the negative of the Hopf algebra Θ).

DEFINITION 1.23. Supposing Θ is an algebra graded on $\mathbb{S}$, and M and M' a pair of right Θ-modules, then $M \otimes M'$ is a module over $\Theta \otimes \Theta$ as in (1.20); but if Θ is also a bialgebra then Θ can be made to act on $M \otimes M'$ through the coproduct $\psi : \Theta \to \Theta \otimes \Theta$. We refer to this action as the diagonal action of Θ, and the resulting Θ-module $M \otimes M'$ as the tensor product of the Θ-modules M and M'.

As in Proposition 2.1 of [**4**] we observe:

PROPOSITION 1.24. *Let Θ be a Hopf algebra, and let M and M' be a pair of right Θ-modules. If M' is a free Θ-module, then $M \otimes M'$ with diagonal Θ-action is isomorphic to a free Θ-module.*

PROOF. We recall the proof of this standard result, because we will need to extend it later. It suffices to consider the case $M' = \Theta$. So it suffices to show that $M \otimes \Theta$ with diagonal action of Θ is isomorphic to $\mathcal{F}^{\Theta}(M)$ of Definition 1.15. The l-linear map $M \otimes \Theta \to M \otimes \Theta$ given by the composition

$$M \otimes \Theta \xrightarrow{M\otimes\psi} M \otimes \Theta \otimes \Theta \xrightarrow{\sigma\otimes\Theta} M \otimes \Theta$$

is Θ-linear when interpreted as a homomorphism $\mathcal{F}^{\Theta}(M) \to M\otimes\Theta$. It has a two-sided inverse given by the composition:

$$M \otimes \Theta \xrightarrow{M\otimes\psi} M \otimes \Theta \otimes \Theta \xrightarrow{M\otimes\chi\otimes\Theta} M \otimes \Theta \otimes \Theta \xrightarrow{\sigma\otimes\Theta} M \otimes \Theta.$$

$\square$

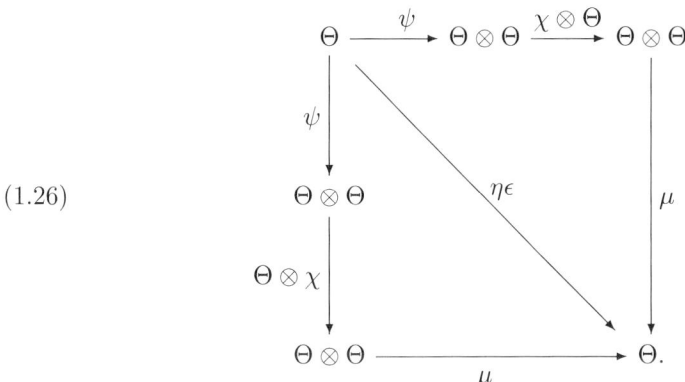

The possibility of forming a tensor product of Θ-modules leads us to pairings of homomorphism groups that extend the pairings ν and $\overline{\nu}$ of (1.23).

DEFINITION 1.25. Let Θ be a bialgebra; suppose M, M', N, N' are right Θ-modules. We will define degree-preserving homomorphisms of vector spaces:

$$\text{(1.29a)} \quad \rho : \mathrm{Hom}_\Theta(M,N) \otimes \mathrm{Hom}_\Theta(M',N') \quad \to \quad \mathrm{Hom}_\Theta(M \otimes M', N \otimes N')$$

$$\text{(1.29b)} \quad \overline{\rho} : \overline{\mathrm{Hom}}^\Theta(M,N) \otimes \overline{\mathrm{Hom}}^\Theta(M',N') \quad \to \quad \overline{\mathrm{Hom}}^\Theta(M \otimes M', N \otimes N').$$

(1.29a) is defined by the diagram

$$\text{(1.30)} \qquad \begin{array}{ccc} \mathrm{Hom}_\Theta(M,N) \otimes \mathrm{Hom}_\Theta(M',N') & & \\ \downarrow \nu & \searrow \rho & \\ \mathrm{Hom}_{\Theta\otimes\Theta}(M \otimes M', N \otimes N') & \xrightarrow[\psi^*]{} & \mathrm{Hom}_\Theta(M \otimes M', N \otimes N') \end{array}$$

where ψ^* is induced by the coproduct $\psi : \Theta \to \Theta \otimes \Theta$. (1.29b) is defined by the same diagram, with Hom replaced everywhere by $\overline{\mathrm{Hom}}$, and ν by $\overline{\nu}$.

The pairings (1.29) will be used in Chapters 2 and 3.

We will often be dealing with modules over Θ that have additional structure.

DEFINITION 1.26. Let Θ be a bialgebra.

1. By a Θ-coalgebra we mean a (commutative) coalgebra Υ that is at the same time a Θ-module, in such a way that the counit $\epsilon : \Upsilon \to l$ and the coproduct $\psi : \Upsilon \to \Upsilon \otimes \Upsilon$ are homomorphisms of Θ-modules.
2. By a Θ-algebra we mean an algebra Ξ that is also a Θ-module, in such a way that the structure maps $\eta : l \to \Xi$ and $\mu : \Xi \otimes \Xi \to \Xi$ are homomorphisms of Θ modules.
3. By a Θ-bialgebra we mean a bialgebra Ξ, that is also a Θ-module, in such a way that the structure maps $\eta : l \to \Xi$, $\mu : \Xi \otimes \Xi \to \Xi$, $\epsilon : \Xi \to l$, $\psi : \Xi \to \Xi \otimes \Xi$, are all homomorphisms of Θ-modules.
4. By a Θ-Hopf algebra we mean a Hopf algebra Ξ, that is also a Θ-module, in such a way that the structure maps $\eta : l \to \Xi$, $\mu : \Xi \otimes \Xi \to \Xi$, $\epsilon : \Xi \to l$, $\psi : \Xi \to \Xi \otimes \Xi$, $\chi : \Xi \to \Xi$ are all homomorphisms of Θ-modules.

We will sometimes refer to a Θ-coalgebra as a "coalgebra over Θ", and will use similar paraphrases for "Θ-algebra", "Θ-bialgebra", and "Θ-Hopf algebra".

PROPOSITION 1.27. *Suppose $\beta : \Theta \to \Delta$ is a homomorphism of bialgebras, and suppose Υ a Δ-coalgebra. Then Υ becomes a Θ-coalgebra, under the action of Θ that is induced by β. Similarly, the pullback under β of a Δ-algebra is a Θ-algebra, the pullback of a Δ-bialgebra is a Θ-bialgebra, and the pullback of a Δ-Hopf algebra is a Θ-Hopf algebra.*

PROPOSITION 1.28. *Let Θ and Θ' be bialgebras. If Υ and Υ' are coalgebras over Θ and Θ' respectively, then the tensor product $\Upsilon \otimes \Upsilon'$ of coalgebras, is a coalgebra over $\Theta \otimes \Theta'$. Similarly, the tensor product of algebras over Θ and Θ' is an algebra over $\Theta \otimes \Theta'$; the tensor product of bialgebras over Θ and Θ' is a bialgebra over $\Theta \otimes \Theta'$, and the tensor product of Hopf algebras over Θ and Θ' is a Hopf algebra over $\Theta \otimes \Theta'$.*

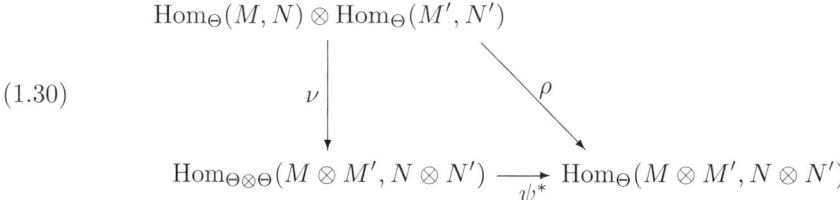

The proofs are straightforward diagram chases. As a corollary we have:

PROPOSITION 1.29. *Let Θ be a bialgebra. If Υ and Υ' are Θ-coalgsebras, then so is the tensor product $\Upsilon \otimes \Upsilon'$ of coalgebras, with diagonal Θ-action. Similarly, the tensor product of Θ-algebras is a Θ-algebra, the tensor product of Θ-bialgebras is a Θ-bialgebra, and the tensor product of Θ-Hopf algebras is a Θ-Hopf algebra.*

This follows from the preceding two propositions. We must use our global assumption that all comultiplications are commutative, so that in particular the coproduct $\psi : \Theta \to \Theta \otimes \Theta$ is a homomorphism of bialgebras.

PROPOSITION 1.30. *Let Θ be a bialgebra over ground field $l = \mathbb{Z}/2$, and Υ a Θ-coalgebra. Then the following diagram commutes:*

$$\begin{array}{ccc} \Upsilon \otimes \Theta & \xrightarrow{\sigma_\Upsilon} & \Upsilon \\ {\scriptstyle V \otimes V} \downarrow & & \downarrow {\scriptstyle V} \\ \Upsilon \otimes \Theta & \xrightarrow[\sigma_\Upsilon]{} & \Upsilon . \end{array}$$

PROOF. The requirement that the coproduct on Υ be a morphism of Θ-modules can also be interpreted as saying that the action of Θ upon Υ is a morphism of coalgebras. Hence our result follows from (1.19) and the naturality of the Verschiebung with respect to coalgebra morphisms. □

Returning to the case of an arbitrary ground field l, we study the duals of Θ-coalgebras.

PROPOSITION 1.31. *Suppose Θ a bialgebra and Υ a left Θ-coalgebra. Let the dual vector space $\overline{\mathrm{Hom}}(\Upsilon, l)$ be considered a right Θ-module, as in (1.25). Then $\overline{\mathrm{Hom}}(\Upsilon, l)$ becomes a right Θ-algebra with product defined as the composition:*

$$(1.31) \qquad \overline{\mathrm{Hom}}(\Upsilon, l) \otimes \overline{\mathrm{Hom}}(\Upsilon, l) \xrightarrow{\overline{\nu}} \overline{\mathrm{Hom}}(\Upsilon \otimes \Upsilon, l) \xrightarrow{\psi_\Upsilon^*} \overline{\mathrm{Hom}}(\Upsilon, l).$$

Here the pairing $\overline{\nu}$ is as in (1.7b).

PROOF. We must show that (1.31) is a homomorphism of right Θ-modules, with Θ acting diagonally on $\overline{\mathrm{Hom}}(\Upsilon, l) \otimes \overline{\mathrm{Hom}}(\Upsilon, l)$. We begin by regarding the vector spaces $\overline{\mathrm{Hom}}(\Upsilon, l) \otimes \overline{\mathrm{Hom}}(\Upsilon, l)$ and $\overline{\mathrm{Hom}}(\Upsilon \otimes \Upsilon, l)$ as modules over $\Theta \otimes \Theta$. Then $\overline{\nu}$ is linear over $\Theta \otimes \Theta$ by Proposition 1.22. So $\overline{\nu}$ is also linear over Θ, where Θ is acting on both $\overline{\mathrm{Hom}}(\Upsilon, l) \otimes \overline{\mathrm{Hom}}(\Upsilon, l)$, and $\overline{\mathrm{Hom}}(\Upsilon \otimes \Upsilon, l)$ through its coproduct. Then the Θ-linearity of $\psi_\Upsilon : \Upsilon \to \Upsilon \otimes \Upsilon$ implies that of ψ_Υ^*, so we are done. □

5. $\Theta - \Xi$ Modules and Relative Homological Algebra

In the same spirit as Definition 1.26 we have:

DEFINITION 1.32. Let Θ be a bialgebra, and Ξ a Θ-algebra. By a $\Theta - \Xi$ module we mean a Ξ-module M that is also a Θ-module, in such a way that the structure map $M \otimes \Xi \to M$ is a homomorphism of Θ-modules. By a homomorphism $f : M \to M'$ from a $\Theta - \Xi$ module M to a $\Theta - \Xi$ module M' and we mean a degree-preserving map that is both a homomorphism of Θ-modules and a homomorphism of Ξ-modules.

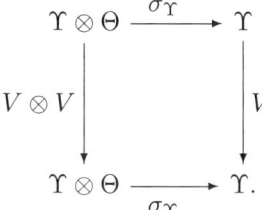

In this section we will establish some simple properties of $\Theta - \Xi$ modules, and show there are enough projectives.

We remark that the definition would make sense even without the requirement that the multiplication on Ξ be Θ-linear. We impose this condition nevertheless: it will play an important role when, later in this section, we do homological algebra in the category of $\Theta - \Xi$ modules. Note we are forming tensor products $M \otimes \Xi$, $\Xi \otimes M$ only when Θ is acting on the same sides of both Ξ and M. So we will have either right $\Theta - \Xi$ modules or left $\Theta - \Xi$ modules, depending on the side from which Θ is acting. However, Ξ can act on M from either side, as in the following:

PROPOSITION 1.33. *Let Θ be a bialgebra and Ξ a right Θ-algebra. Let M and N be a pair of right $\Theta - \Xi$ modules, such that Ξ acts on the right of M and on the left of N. Then the diagonal action of Θ on $M \otimes N$ passes to an action on the quotient $M \otimes_\Xi N$, so that $M \otimes_\Xi N$ becomes a right Θ-module.*

This is a straightforward consequence of the definitions. Another is that the property of being a $\Theta - \Xi$ module can be "pulled back" along a bialgebra homomorphism of the variable Θ:

PROPOSITION 1.34. *Let Δ be a bialgebra, Ξ a right Δ-algebra, and M a $\Delta - \Xi$ module. Suppose given a homomorphism of bialgebras $\beta : \Theta \to \Delta$. Then with Θ acting on the right of both Ξ and M through β, Ξ becomes a right Θ-algebra, and M becomes a $\Theta - \Xi$ module.*

The first part is already in Proposition 1.27, and the second is easy. One can also pull back along the variable Ξ:

PROPOSITION 1.35. *Let Θ be a bialgebra, Υ a right Θ algebra, and M a $\Theta - \Upsilon$ module. Suppose given a homomorphism $\beta : \Xi \to \Upsilon$ of Θ-algebras. Then with Ξ acting on M through β, M becomes a $\Theta - \Xi$ module.*

We have further:

PROPOSITION 1.36. *Suppose given:*

1. *A bialgebra Θ, a right Θ-algebra Ξ, and a $\Theta - \Xi$ module M on which Ξ acts from the right;*
2. *A bialgebra Θ', a right Θ'-algebra Ξ', and a $\Theta' - \Xi'$ module M' on which Ξ' acts from the right.*

Then $\Xi \otimes \Xi'$ is a right algebra over the bialgebra $\Theta \otimes \Theta'$, and $M \otimes M'$ is a $\Theta \otimes \Theta' - \Xi \otimes \Xi'$ module.

The first part is already in Proposition 1.28, and the second is a straightforward diagram chase. As a corollary we have:

PROPOSITION 1.37. *Suppose Θ a bialgebra and Ξ a right Θ-bialgebra. Suppose M and M' are $\Theta - \Xi$ modules, with Ξ acting on the same side of M and M'. Then $M \otimes M'$ is also a $\Theta - \Xi$ module, with diagonal actions of both Θ and Ξ.*

PROOF. By Proposition 1.36 we have that $M \otimes M'$ is a $\Theta \otimes \Theta - \Xi \otimes \Xi$ module. As in Proposition 1.34 we pull back this structure along the diagonal $\psi_\Theta : \Theta \to \Theta \otimes \Theta$ (a bialgebra homomorphism by the cocommutativity of Θ), making $M \otimes M'$ into a $\Theta - (\Xi \otimes \Xi)$ module. Now as in Proposition 1.35 we pull back this new structure along the diagonal $\psi_\Xi : \Xi \to \Xi \otimes \Xi$, a homomorphism of Θ-algebras. This last pullback completes the proof. □

We will refer to $M \otimes M'$ with diagonal actions of both Θ and Ξ as the tensor product of the $\Theta - \Xi$ modules M and M'.

One can also mix free and diagonal actions. In the following proposition, $\mathcal{F}^{\Xi}(M)$ is the free right Ξ-module generated by M, as in Definition 1.15.

PROPOSITION 1.38. *Suppose Θ a bialgebra, and Ξ a Θ-algebra. Suppose M a $\Theta - \Xi$ module, with Ξ acting from the right. Then $\mathcal{F}^{\Xi}(M)$ becomes a $\Theta - \Xi$ module if we give it the diagonal Θ-action.*

This is an immediate consequence of the assumption that the multiplication $\Xi \otimes \Xi \to \Xi$ is a homomorphism of Θ-modules. We will write $\mathcal{F}^{\Xi}(M)$ for the $\Theta - \Xi$ module described in Proposition 1.38.

Of course the above proposition would be valid if M were only a Θ-module. But we will be using the proposition only in cases for which M has actions of both Θ and Ξ.

PROPOSITION 1.39. *Let Θ be a bialgebra, Ξ a Θ-Hopf algebra, and M a $\Theta - \Xi$ module with Ξ acting from the right. Then $\mathcal{F}^{\Xi}(M)$ is isomorphic to the tensor product of the $\Theta - \Xi$ modules M and Ξ described in Proposition 1.37.*

PROOF. The isomorphism $\mathcal{F}^{\Xi}(M) \to M \otimes \Xi$ given in the proof of Proposition 1.24 is in the present context a mapping of $\Theta - \Xi$ modules. The key point is that the coproduct on Ξ is Θ-linear by hypothesis. □

In Chapter 3, and in our treatments of the change-of-rings and Eilenberg-Moore spectral sequences that depend on that chapter, we will need to do some relative homological algebra in the category of $\Theta - \Xi$ modules. We review some basic definitions [**21**],[**28**].

DEFINITION 1.40. Let Θ be a bialgebra and Ξ a right Θ-algebra. An epimorphism $k : T \to T'$ of right $\Theta - \Xi$ modules is called relatively split if k has a right inverse $j : T' \to T$ in the category of Θ-modules.

DEFINITION 1.41. Let Θ be a bialgebra and Ξ a right Θ-algebra. A right $\Theta - \Xi$ module $\mathcal{P}$ is called a relative projective if it is projective as a Ξ-module, and in addition satisfies the following condition. Suppose given any diagram of the form

(1.32)

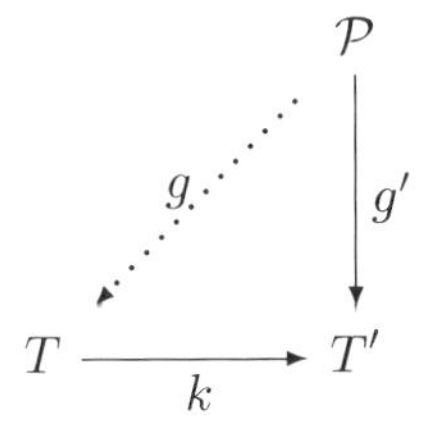

where g' and k are homomorphisms of right $\Theta - \Xi$-modules and k is a relatively split epimorphism. Then there exists a homomorphism $g : \mathcal{P} \to T$ of $\Theta - \Xi$ modules such that $kg = g'$.

PROPOSITION 1.42. *Let Θ be a bialgebra and Ξ a right Θ-algebra. Suppose M any right $\Theta - \Xi$ module. Then the $\Theta - \Xi$ module $\mathcal{F}^{\Xi}(M)$ of Proposition 1.38 is a relative projective.*

PROOF. Suppose given a diagram:

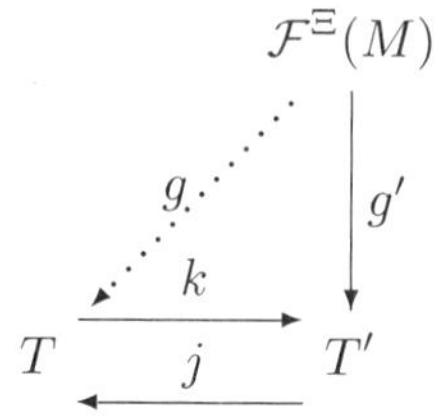

Here g' is a homomorphism of $\Theta - \Xi$ modules, k is a relatively split epimorphism of $\Theta - \Xi$ modules and j is a Θ-linear right inverse for k. The lift g is to be constructed: a homomorphism of $\Theta - \Xi$ modules for which $kg = g'$. We identify $\mathcal{F}^{\Xi}(M)$ with $M \otimes \Xi$ as vector spaces over l, and begin by observing that since Ξ is a Θ-algebra, the unit $l \to \Xi$ is a homomorphism of Θ-modules; consequently, $M \equiv M \otimes l$ is a Θ-submodule of $\mathcal{F}^{\Xi}(M)$. Write $g'|M$ for the restriction of g' to this submodule. Define a Θ-linear $h : M \to T$ by setting $h = j(g'|M)$. Define g to be the unique Ξ-linear extension of h to $\mathcal{F}^{\Xi}(M)$. This is the composition: $M \otimes \Xi \xrightarrow{h \otimes \Xi} T \otimes \Xi \xrightarrow{\sigma_T} T$. Since σ_T is Θ-linear, so is g. The relation $kg = g'$ is clearly satisfied. As the action σ_T is a homomorphism of degree zero, so g also has degree zero, as required. □

PROPOSITION 1.43. *Let Θ be a bialgebra and Ξ a right Θ-Hopf algebra. Suppose M is any right $\Theta - \Xi$ module. Then the tensor product $M \otimes \Xi$ of the $\Theta - \Xi$ modules M and Ξ is a relatively projective $\Theta - \Xi$ module.*

PROOF. This follows at once from Propositions 1.39 and 1.42. □

Another corollary of Proposition 1.42 is that the category of $\Theta - \Xi$ modules has enough relative projectives:

PROPOSITION 1.44. *Let Θ be a bialgebra and Ξ a right Θ-algebra. Suppose M any right $\Theta - \Xi$ module. Then there exists a relatively projective $\Theta - \Xi$ module mapping onto M by a relatively split epimorphism.*

PROOF. We identify $\mathcal{F}^{\Xi}(M)$ with $M \otimes \Xi$ as vector spaces over l. The action of Ξ is on the right hand factor and the action of Θ on $M \otimes \Xi$ is diagonal. The mapping $\mathcal{F}^{\Xi}(M) \equiv M \otimes \Xi \xrightarrow{\sigma_M} M$ is linear over both Θ and Ξ and has degree zero. We will show this is a relatively split epimorphism. Since the unit $\eta : l \to \Xi$ is Θ-linear, the mapping $M \equiv M \otimes l \xrightarrow{M \otimes \eta} M \otimes \Xi = \mathcal{F}^{\Xi}(M)$ is also Θ-linear (although it is certainly not Ξ-linear!), and provides a relative splitting for σ_M. □

DEFINITION 1.45. Suppose Θ a bialgebra, Ξ a right Θ-algebra, and M a right $\Theta - \Xi$ module. By a resolution of M by relative projectives we mean a chain complex of relatively projective $\Theta - \Xi$ modules, augmenting to M, in which each differential, and the augmentation, are homomorphisms of $\Theta - \Xi$ modules. We further require that this augmented complex has a contracting homotopy that is Θ-linear. We will write $\mathcal{P}(M)$ for the complex: $\cdots \to \mathcal{P}_p(M) \to \cdots \to \mathcal{P}_0(M)$, and $\mathcal{P}(M) \to M \to 0$ for the augmented version.

REMARK 1.46. Proposition 1.44 implies, through standard arguments of homological algebra, that every $\Theta - \Xi$ module M has a resolution by relative projectives; and further, that such a resolution is unique up to chain-homotopy equivalence of complexes of $\Theta - \Xi$ modules.

6. Algebras with Coproducts

In his classic paper [**52**], Milnor showed that the Steenrod algebra $\mathcal{A}$ that acts on the mod-2 cohomology of topological spaces is a Hopf algebra. But there is another version of the Steenrod algebra that we call $\mathcal{H}$ whose definition will be reviewed in Section 12. While in $\mathcal{A}$ one has $Sq^0 = 1$, in $\mathcal{H}$ the situation is quite different: Sq^0 is an independent generator. We emphasize three consequences of this fact. The first is that $\mathcal{H}$ is bigraded as an algebra ... graded on total degree, like $\mathcal{A}$, but also on the length of a monomial in the generating set $\{Sq^k \mid k \geq 0\}$. The second is that $\mathcal{H}$ is not a Hopf algebra ... one cannot define $\chi(Sq^0)$. The third is that $\mathcal{H}$ is not even a bialgebra; at least, not in the sense we have defined, since the coproduct does not preserve bidegree. So we have to introduce a class of objects called algebras with coproducts, generalizing the bialgebras, of which $\mathcal{H}$ will be an example, and establish for them analogues of the definitions and theorems of the previous section. A reader interested only in the Eilenberg-Moore spectral sequence will need only the definition of an algebra with coproducts. But the reader interested in the change-of-rings spectral sequence for an extension of Hopf algebras should read all of this section. Its results will be used in our description of the vertical Steenrod operations on the E_2-term.

Although $\mathcal{H}$ is the only algebra with coproducts that will appear in this work that is not a Hopf algebra, the Dyer-Lashof algebra is also such an object. The notation and language we are about to introduce may be useful in discussions of that case as well.

In this section we assume that our ground field l has characteristic $p > 0$.

DEFINITION 1.47. Let $\mathbb{T}$ be an abelian group, and $\rho' : \mathbb{T} \to \mathbb{Z}$ a homomorphism. By an algebra with coproducts graded on $\mathbb{T} \times \mathbb{Z}$ we mean, first of all, a $\mathbb{T} \times \mathbb{Z}$ - graded algebra $\Pi = \{\Pi_{t,v} \mid (t,v) \in \mathbb{T} \times \mathbb{Z}\}$. An element in $\Pi_{t,v}$ will be said to have additive degree t and multiplicative degree v, and total degree will be defined by $\rho(t,v) = \rho'(t)$. For each $v \in \mathbb{Z}$ we write $\Pi_{*,v} = \{\Pi_{t,v} \mid t \in \mathbb{T}\}$ for the $\mathbb{T}$-graded vector space obtained by fixing the multiplicative degree, and impose the additional requirements:

1. For each $v \in \mathbb{Z}$, $\Pi_{*,v}$ is given the structure of a commutative $\mathbb{T}$-graded coalgebra;
2. The algebra unit η, regarded as a mapping $\eta : l \to \Pi_{*,0}$ is a morphism of coalgebras;
3. For all $(u,v) \in \mathbb{Z} \times \mathbb{Z}$, the product $\mu : \Pi_{*,u} \otimes \Pi_{*,v} \to \Pi_{*,u+v}$ is a morphism of $\mathbb{T}$-graded coalgebras.

An ordinary $\mathbb{T}$-graded bialgebra Π can be considered an algebra with coproducts concentrated in multiplicative degree 0.

A homomorphism of algebras with coproducts is required to preserve both additive and multiplicative degrees, and to preserve in the obvious sense all structures listed in Definition 1.47.

We define the cross product $\Pi \times \Pi'$ of two algebras with coproducts, assumed graded on the same abelian group $\mathbb{T} \times \mathbb{Z}$. For each $v \in \mathbb{Z}$ the coalgebra $(\Pi \times \Pi')_{*,v}$ is by definition the tensor product of $\mathbb{T}$-graded coalgebras:

$$(\Pi \times \Pi')_{*,v} = \Pi_{*,v} \otimes \Pi'_{*,v}. \tag{1.33}$$

For each pair $u, v \in \mathbb{Z}$ we define a homomorphism of vector spaces:

$$(\Pi \times \Pi')_{*,u} \otimes (\Pi \times \Pi')_{*,v} \to (\Pi \times \Pi')_{*,u+v}$$

to be the following composition:

$$(1.34) \quad (\Pi_{*,u} \otimes \Pi'_{*,u}) \otimes (\Pi_{*,v} \otimes \Pi'_{*,v}) \xrightarrow{(1,3,2,4)} (\Pi_{*,u} \otimes \Pi_{*,v}) \otimes (\Pi'_{*,u} \otimes \Pi'_{*,v}) \text{—} \xrightarrow{\mu_\Pi \otimes \mu_{\Pi'}} \Pi_{*,u+v} \otimes \Pi'_{*,u+v}.$$

This composition is clearly a homomorphism of coalgebras for each pair $u, v \in \mathbb{Z}$. The collection of these mappings for all $u, v \in \mathbb{Z}$ gives $\Pi \times \Pi'$ the structure of a $(\mathbb{T} \times \mathbb{Z})$-graded algebra, so, under the definitions (1.33),(1.34), our cross product $\Pi \times \Pi'$ becomes an algebra with coproducts in its own right.

If both Π and Π' are concentrated in multiplicative degree 0 then they are bialgebras, and the cross product $\Pi \times \Pi'$ reduces to the ordinary tensor product $\Pi \otimes \Pi'$ of bialgebras.

REMARK 1.48. If Π is an algebra with coproducts then the coproducts

$$\psi : \Pi_{*,v} \to \Pi_{*,v} \otimes \Pi_{*,v}$$

define a homomorphism $\psi : \Pi \to \Pi \times \Pi$ of algebras with coproducts. (Of course we are using here our assumption that coproducts are commutative.)

We have required that the actions of ordinary algebras and bialgebras on their modules be degree-preserving. However, algebras with coproducts act on their modules in a special way. We will give separate rules for left modules and for right. These rules may appear somewhat arbitrary now, but the justification will appear in Chapter 3, Proposition 3.22, when we define the action of algebras with coproducts on homomorphism groups. We remind the reader of our Definition 1.3 of uncontracted tensor products $V \widehat{\otimes} W$, and that our ground field has characteristic $p > 0$.

DEFINITION 1.49. Let Π be an algebra with coproducts graded on $\mathbb{T} \times \mathbb{Z}$, let $\mathbb{S}$ be an abelian group, and suppose M a vector space graded on $\mathbb{T} \times \mathbb{S}$. We say M is a right Π-module if there is given for each $v \in \mathbb{Z}$ and each $s \in \mathbb{S}$ a degree-preserving homomorphism of $\mathbb{T}$-graded vector spaces $M_{*,s} \otimes \Pi_{*,v} \to M_{*,s/p^v}$, in such a way that the associated mapping $\sigma : M \widehat{\otimes} \Pi \to M$ satisfies the unitary and associative laws, and is parity-preserving in the sense of Definitions 1.3 and 1.6. Elements of $M_{t,s}$, with $(t, s) \in \mathbb{T} \times \mathbb{S}$, are said to have additive degree t and multiplicative degree s.

We interpret $M_{*,s/p^v}$ to mean the zero vector space if s is not divisible by p^v in $\mathbb{S}$. The indexing groups that we use are torsion free, so that expressions like s/p^v are unambiguous when defined. The "special modules" of Definition 5.10 will be our main examples of right modules over an algebra with coproducts.

Notice that we cannot define a vector space $M \otimes \Pi$ in the present context, since M and Π are not necessarily graded on the same group. So we are not allowed to describe the action of Π on M as a homomorphism $M \otimes \Pi \to M$. Even in the case $\mathbb{S} = \mathbb{Z}$, when $M \otimes \Pi$ would be defined as a $\mathbb{T} \times \mathbb{Z}$-graded vector space, the action of Π on M is not a mapping $M \otimes \Pi \to M$ because Definition 1.5 would not be satisfied... not all elements of equal degree in $M \otimes \Pi$ are mapped to the same degree in M.

Nevertheless we can describe the action of Π as a mapping $\sigma : M\widehat{\otimes}\Pi \to M$; and since the multiplication $\mu : \Pi \otimes \Pi \to \Pi$ can also be interpreted as a homomorphism $\Pi\widehat{\otimes}\Pi \to \Pi$, the requirement of associativity in the above definition can, by analogy with the classical case, be expressed as commutativity of a diagram of vector space homomorphisms:

$$\begin{array}{ccc} M\widehat{\otimes}\Pi\widehat{\otimes}\Pi & \xrightarrow{M \otimes \mu} & M\widehat{\otimes}\Pi \\ {\scriptstyle \sigma \otimes \Pi}\downarrow & & \downarrow{\scriptstyle \sigma} \\ M\widehat{\otimes}\Pi & \xrightarrow[\sigma]{} & M. \end{array}$$

DEFINITION 1.50. Let Π be an algebra with coproducts graded on $\mathbb{T} \times \mathbb{Z}$, let $\mathbb{S}$ be an abelian group, and suppose N is a vector space graded on $\mathbb{T} \times \mathbb{S}$. We say N is a left Π-module if there is given for each $v \in \mathbb{Z}$ and each $s \in \mathbb{S}$ a degree-preserving homomorphism of $\mathbb{T}$-graded vector spaces $\Pi_{*,v} \otimes N^{*,s} \to N^{*,p^v s}$, in such a way that the associated mapping $\sigma : \Pi\widehat{\otimes}N \to N$ satisfies the unitary and associative laws, and is parity-preserving in the sense of Definitions 1.3 and 1.6. Elements of $N^{t,s}$, with $(t, s) \in \mathbb{T} \times \mathbb{S}$, are said to have additive degree t and multiplicative degree s.

We will see in Section 12 that the cohomology ring of a graded Hopf algebra over $\mathbb{Z}/2$ is a left module over the algebra with coproducts $\boldsymbol{\mathcal{H}}$ of Steenrod operations.

DEFINITION 1.51. Let Π be an algebra with coproducts graded on $\mathbb{T} \times \mathbb{Z}$, and M, M' a pair of either left or right Π-modules graded on the same group $\mathbb{T} \times \mathbb{S}$. By a homomorphism f of Π-modules from M to M' we mean a degree-preserving homomorphism of $(\mathbb{T} \times \mathbb{S})$-graded vector spaces $f : M \to M'$ that commutes with Π action.

Suppose Π and Π' are algebras with coproducts graded on the same group $\mathbb{T} \times \mathbb{Z}$, and that M and M' are right modules over Π and Π', respectively, graded on the same group $\mathbb{T} \times \mathbb{S}$. For each pair $(i, j) \in \mathbb{S} \times \mathbb{S}$ and each $v \in \mathbb{Z}$ we define a homomorphism of $\mathbb{T}$-graded vector spaces: $\sigma : (M)_{*,i} \otimes (M')_{*,j} \otimes (\Pi \times \Pi')_{*,v} \to (M \otimes M')_{*,(i+j)/p^v}$ to be the composition:

(1.35)

$$(M)_{*,i} \otimes (M')_{*,j} \otimes \Pi_{*,v} \otimes \Pi'_{*,v} \xrightarrow{(1,3,2,4)} (M)_{*,i} \otimes \Pi_{*,v} \otimes (M')_{*,j} \otimes \Pi'_{*,v} \text{---}$$
$$\xrightarrow{\sigma_M \otimes \sigma_{M'}} (M)_{*,i/p^v} \otimes (M')_{*,j/p^v}.$$

For fixed $s \in \mathbb{S}, v \in \mathbb{Z}$ we then take the direct sum of the homomorphisms (1.35) over all pairs (i, j) for which $i + j = s$, obtaining $(M \otimes M')_{*,s} \otimes (\Pi \times \Pi')_{*,v} \to (M \otimes M')_{*,s/p^v}$.

DEFINITION 1.52. Suppose that Π and Π' are algebras with coproducts graded on the same group $\mathbb{T} \times \mathbb{Z}$, and that M and M' are right modules over Π and Π', respectively, graded on the same group $\mathbb{T} \times \mathbb{S}$. By the tensor product of M and M'

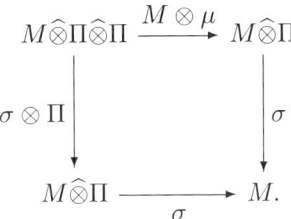

we mean the right $(\Pi \times \Pi')-$module whose underlying $\mathbb{T} \times \mathbb{S}$-graded vector space is $M \otimes M'$, and for which the action of $\Pi \times \Pi'$ is defined by (1.35). The tensor product of left modules over a pair of algebras with coproducts is defined similarly.

DEFINITION 1.53. Suppose Π an algebra with coproducts graded on $\mathbb{T} \times \mathbb{Z}$, and M and M' a pair of right Π-modules graded on the same group $\mathbb{T} \times \mathbb{S}$. Then we have from the previous definition that $M \otimes M'$ is a right module over $\Pi \times \Pi$. But we have observed in Remark 1.48 that the coproduct $\psi : \Pi \to \Pi \times \Pi$ can be regarded as a homomorphism of algebras with coproducts. Consequently we can regard $M \otimes M'$ as a right module over Π, acting through its coproduct. We will refer to this action as the diagonal action of Π, and the resulting module $M \otimes M'$ as the tensor product of the Π-modules M and M'.

Diagonal action of algebras with coproducts on a tensor product of left modules is defined similarly.

REMARK 1.54. Now that we have a notion of a tensor product of modules over an algebra with coproducts Π, we can easily write down the definitions of a Π-coalgebra, Π-algebra, Π-bialgebra, Π-Hopf algebra. These are formally identical to the corresponding definitions that we gave in Definition 1.26, for the case in which Π was a bialgebra.

By analogy with Propositions 1.28 and 1.29 we have:

PROPOSITION 1.55. *Let Π and Π' be algebras with coproducts graded on the same group $\mathbb{T} \times \mathbb{Z}$. Suppose Υ and Υ' are coalgebras over Π and Π' respectively, both graded on $\mathbb{T} \times \mathbb{S}$. Then the tensor product $\Upsilon \otimes \Upsilon'$ of coalgebras, is a coalgebra over $\Pi \times \Pi'$. Similarly, the tensor product of algebras over Π and Π' is an algebra over $\Pi \times \Pi'$; the tensor product of bialgebras over Π and Π' is a bialgebra over $\Pi \times \Pi'$, and the tensor product of Hopf algebras over Π and Π' is a Hopf algebra over $\Pi \times \Pi'$.*

PROPOSITION 1.56. *Let Π be an algebra with coproducts. If Υ and Υ' are Π-coalgebras, then so is the tensor product $\Upsilon \otimes \Upsilon'$ of coalgebras, with diagonal Π-action. Similarly, the tensor product of Π-algebras is a Π-algebra, the tensor product of Π-bialgebras is a Π-bialgebra, and the tensor product of Π-Hopf algebras is a Π-Hopf algebra.*

Of course, Proposition 1.56 follows from Proposition 1.55 in the same way that Proposition 1.29 follows from Proposition 1.28. Structures are pulled back along the diagonal $\psi : \Pi \to \Pi \times \Pi$, which is a homomorphism of algebras with coproducts.

We have by analogy with Definition 1.32:

DEFINITION 1.57. Let Π be an algebra with coproducts, and suppose Ξ a right Π-algebra. Then by a right $\Pi - \Xi$ module we mean a Ξ-module M, that is also a right Π-module, in such a way that the action $M \otimes \Xi \to M$ (or the action $\Xi \otimes M \to M$) is a homomorphism of right Π-modules. By a homomorphism $f : M \to M'$ from a right $\Pi - \Xi$ module M to a right $\Pi - \Xi$ module M' and we mean a map that preserves both additive and multiplicative degrees, and that is both a homomorphism of Π-modules and a homomorphism of Ξ-modules.

The following generalization of Proposition 1.38 provides examples of $\Pi - \Xi$ modules.

PROPOSITION 1.58. *Suppose* Π *is an algebra with coproducts, and suppose* Ξ *is a right* Π*-algebra. Suppose* M *is a right* $\Pi - \Xi$ *module, with* Ξ *acting from the right. Then* $\mathcal{F}^{\Xi}(M)$ *becomes a* $\Pi - \Xi$ *module if given the diagonal* Π*-action.*

REMARK 1.59. If in Propositions 1.34 and 1.35 we regard Θ and Δ as algebras with coproducts, rather than bialgebras, the propositions remain true. That is, the property of being a $\Pi - \Xi$ module can be pulled back along homomorphisms of the variable Π, and along Π-algebra homomorphisms of the variable Ξ.

By analogy with Proposition 1.36 we have:

PROPOSITION 1.60. *Suppose given:*

1. *An algebra with coproducts* Π*, a right* Π*-algebra* Ξ*, and a* $\Pi - \Xi$ *module* M *on which* Ξ *acts from the right;*
2. *An algebra with coproducts* Π'*, a right* Π'*-algebra* Ξ'*, and a* $\Pi' - \Xi'$ *module* M' *on which* Ξ' *acts from the right.*

Then $\Xi \otimes \Xi'$ *is a right algebra over the algebra with coproducts* $\Pi \times \Pi'$*, and* $M \otimes M'$ *is a* $\Pi \times \Pi' - \Xi \otimes \Xi'$ *module.*

Then by analogy with Proposition 1.37:

PROPOSITION 1.61. *Suppose* Π *is an algebra with coproducts and suppose* Ξ *is a right* Π*-bialgebra. Suppose* M *and* M' *are* $\Pi - \Xi$ *modules, with* Ξ *acting on the same side of* M *and* M'*. Then* $M \otimes M'$ *is also a* $\Pi - \Xi$ *module, with diagonal actions of both* Π *and* Ξ*.*

This follows from Proposition 1.60 and Remark 1.59 in the same way that Proposition 1.37 follows from Propositions 1.34, 1.35, and 1.36.

We will refer to $M \otimes M'$ with diagonal actions of both Π and Ξ as the tensor product of the $\Pi - \Xi$ modules M and M'.

Now we can generalize Proposition 1.39:

PROPOSITION 1.62. *Let* Π *be an algebra with coproducts,* Ξ *a right* Π*-Hopf algebra, and* M *a* $\Pi - \Xi$ *module with* Ξ *acting from the right. Then* $\mathcal{F}^{\Xi}(M)$ *of Proposition 1.58, and* $M \otimes \Xi$*, are isomorphic* $\Pi - \Xi$ *modules.*

We develop a theory of relatively projective $\Pi - \Xi$ modules, by analogy with the work in Section 5.

DEFINITION 1.63. Let Π be an algebra with coproducts and suppose Ξ is a right Π-algebra. An epimorphism $k : T \to T'$ of right $\Pi - \Xi$ modules is called relatively split if k has a right inverse $j : T' \to T$ in the category of Π-modules.

REMARK 1.64. We define a "relatively projective" $\Pi - \Xi$ module by analogy with Definition 1.41. We replace the bialgebra Θ of that definition with the algebra with coproducts Π. The mapping k of (1.32) is a relatively split epimorphism of $\Pi - \Xi$ modules, and the $\Pi - \Xi$ module $\mathcal{P}$ is called relatively projective if it is projective as a Ξ-module, and if triangles of the form (1.32) can always be filled.

Generalizing Proposition 1.42 we have:

PROPOSITION 1.65. *Let* Π *be an algebra with coproducts and suppose* Ξ *is a right* Π*-algebra. Suppose* M *any right* $\Pi - \Xi$ *module. Then the* $\Pi - \Xi$ *module* $\mathcal{F}^{\Xi}(M)$ *of Proposition 1.58 is a relative projective.*

Then we have from Propositions 1.62 and 1.65:

PROPOSITION 1.66. *Let* Π *be an algebra with coproducts and let* Ξ *be a right* Π-*Hopf algebra. Suppose that* M *is any right* $\Pi - \Xi$ *module. Then the tensor product* $M \otimes \Xi$ *of the* $\Pi - \Xi$ *modules* M *and* Ξ *is a relatively projective* $\Pi - \Xi$ *module.*

Another corollary of Proposition 1.65 is that the category of $\Pi - \Xi$ modules has enough relative projectives. The following proposition generalizes Proposition 1.44 and is proved in the same way.

PROPOSITION 1.67. *Let* Π *be an algebra with coproducts and suppose* Ξ *is a right* Π-*algebra. Suppose* M *is any right* $\Pi - \Xi$ *module. Then there exists a relatively projective* $\Pi - \Xi$ *module mapping onto* M *by a relatively split epimorphism. In fact,* $\mathcal{F}^{\Xi}(M) \equiv M \otimes \Xi \xrightarrow{\sigma_M} M$ *is such a mapping.*

DEFINITION 1.68. Suppose that Π is an algebra with coproducts , Ξ is a right Π-algebra, and M is a right $\Pi - \Xi$ module. By a resolution of M by relative projectives we mean a chain complex of relatively projective $\Pi - \Xi$ modules, augmenting to M, in which the differentials, and the augmentation, are homomorphisms of $\Pi - \Xi$ modules. We further require that this augmented complex have a contracting homotopy that is Π-linear. We will write $\mathcal{P}(M)$ for the complex: $\cdots \to \mathcal{P}_p(M) \to \cdots \to \mathcal{P}_0(M)$, and $\mathcal{P}(M) \to M \to 0$ for the augmented version.

REMARK 1.69. Proposition 1.67 implies, through standard arguments of homological algebra, that every $\Pi - \Xi$ module M has a resolution by relative projectives; and further, that such a resolution is unique up to chain-homotopy equivalence of complexes of $\Pi - \Xi$ modules.

We conclude this section on algebras with coproducts and their modules with a discussion of negatives.

DEFINITION 1.70. Let Π be an algebra with coproducts graded on $\mathbb{T} \times \mathbb{Z}$. By its "negative", written $\overline{\Pi}$, we mean the algebra with coproducts for which the underlying bigraded vector space is defined by $\overline{\Pi}_{t,v} = \Pi_{-t,v}$ for all $(t, v) \in \mathbb{T} \times \mathbb{Z}$. Elements of $\overline{\Pi}_{t,v}$ have additive degree t and multiplicative degree v. Product and coproduct on $\overline{\Pi}$ are induced in the obvious way from product and coproduct on Π.

This definition is a little inconsistent with language we have previously introduced, in that the $(\mathbb{T} \times \mathbb{Z})$-graded vector space $\overline{\Pi}$ that underlies the algebra with coproducts $\overline{\Pi}$ is not the negative of the $(\mathbb{T} \times \mathbb{Z})$-graded vector space Π as given in Definition 1.2. The following remark provides some justification for our definitions. More will be found in the results of Chapter 3, particularly Propostion 3.22.

REMARK 1.71. Suppose M a right module over the algebra with coproducts Π. Then the vector space $\overline{M}$ as given in Definition 1.2 becomes a right module over the algebra with coproducts $\overline{\Pi}$, with the multiplication $\overline{M} \widehat{\otimes} \overline{\Pi} \to \overline{M}$ defined by the given multiplication $M \widehat{\otimes} \Pi \to M$ in the obvious way. Similarly, if N is a left Π module, then the vector space $\overline{N}$ becomes a left module over $\overline{\Pi}$. We will refer to $\overline{M}$ and $\overline{N}$ as the negatives of the Π-modules M and N.

7. Chains and Cochains

Fix an algebra Θ graded on an abelian group $\mathbb{S}$. A chain complex C over the category of Θ-modules is a family $\{C_n \mid n \in \mathbb{Z}\}$ of $\mathbb{S}$-graded Θ-modules with differentials $\cdots C_n \xrightarrow{\partial} C_{n-1} \xrightarrow{\partial} \cdots$ that are homomorphisms of Θ-modules that preserve the $\mathbb{S}$-degree. An element in $C_{n,s}$ for $n \in \mathbb{Z}$ and $s \in \mathbb{S}$ will be said to have homological degree n and internal degree s. We will almost always suppress the internal degree. In some contexts it is useful to consider C as a vector space graded on $\mathbb{Z} \times \mathbb{S}$, with the first factor $\mathbb{Z}$ giving the homological degree. Then we must define the total degree with a homomorphism $\rho' : \mathbb{Z} \times \mathbb{S} \to \mathbb{Z}$ as in Section 1. We choose the formula: $\rho'(n, s) = n + \rho(s)$. For each $n \in \mathbb{Z}$, the n'th homology group of C is itself graded on $\mathbb{S}$:

$$H_nC = \{H_{n,s}C \mid s \in \mathbb{S}\}.$$

The $(\mathbb{Z} \times \mathbb{S})$-graded vector space $H_{*,*}C$ is defined by

$$H_{*,*}C = \{H_{n,s}C \mid n \in \mathbb{Z},\ s \in \mathbb{S}\}.$$

Similarly a cochain complex C over Θ is a cochain complex over the category of Θ-modules, with differentials $\delta : C^n \to C^{n+1}$ that preserve internal degree. When considering C as a vector space graded on $\mathbb{Z} \times \mathbb{S}$, we will define total degree by the formula $\rho'(n, s) = n + \rho(s)$. The n'th homology group of C under the differential δ is itself graded on $\mathbb{S}$:

$$H^nC = \{H^{n,s}C \mid s \in \mathbb{S}\}.$$

The $(\mathbb{Z} \times \mathbb{S})$-graded vector space $H^{*,*}C$ is defined by

$$H^{*,*}C = \{H^{n,s}C \mid n \in \mathbb{Z},\ s \in \mathbb{S}\}. \tag{1.36}$$

Suppose that Θ is a bialgebra and that C and D are chain complexes over the category of Θ-modules. We denote by $C \otimes D$ the usual tensor product of chain complexes, with diagonal Θ-action. In the formula $(C \otimes D)_n = \bigoplus_{p+q=n}(C_p \otimes D_q)$, only homological degrees are displayed, however each term $C_p \otimes D_q$ is a tensor product of $\mathbb{S}$-graded Θ-modules. In the formula $\partial(c \otimes d) = \partial c \otimes d + (-1)^{p+|c|}\, c \otimes \partial d$, the expression $p+|c|$ refers to the total degree of c in the chain complex, as explained above. The tensor product of cochain complexes is defined analogously.

Definition 1.72. Suppose that Θ is a bialgebra and that C and D are chain complexes over the category of Θ-modules. By a homomorphism $f : C \to D$ of chain complexes ("chain map") we mean a degree-preserving homomorphism of $(\mathbb{Z} \times \mathbb{S})$-graded vector spaces that commutes with the differentials in the obvious sense. Homomorphisms of cochain complexes ("cochain maps") are defined similarly.

The following definition is standard.

Definition 1.73. Let Θ be an algebra graded on $\mathbb{S}$, and let C and D be chain complexes over the category of Θ-modules. We will define a cochain complex $\mathrm{Hom}_\Theta(C, D)$ over the category of $\mathbb{S}$-graded vector spaces. For each $n \in \mathbb{Z}$, the $\mathbb{S}$-graded vector space $\mathrm{Hom}^n_\Theta(C, D)$ is defined by

$$\mathrm{Hom}^n_\Theta(C, D) = \prod_{p-q=n} \mathrm{Hom}_\Theta(C_p, D_q), \tag{1.37}$$

where $\mathrm{Hom}_\Theta(C_p, D_q)$ is as in Definitions 1.16 and 1.17. We define the differential $\delta : \mathrm{Hom}^n_\Theta(C, D) \to \mathrm{Hom}^{n+1}_\Theta(C, D)$ by:

$$(\delta f)(c) = \partial(f(c)) + (-1)^{n+|f|} f(\partial c) \tag{1.38}$$

for all $f \in \mathrm{Hom}^n_\Theta(C, D)$ and all $c \in C$. Here $n + |f|$ is the total degree of f in the cochain complex, as explained above.

Equation (1.38) is of course a shorthand, like that described in Remark 1.7.

REMARK 1.74. A homomorphism $f : C \to D$ of chain complexes in the sense of Definition 1.72 is a cocycle in $\mathrm{Hom}_\Theta(C, D)$ with both homological and internal degrees equal to zero.

A special case of the construction in Definition 1.73 occurs when D is just a module N over Θ, regarded as a chain complex concentrated in homological degree 0. Then we get a cochain complex $\mathrm{Hom}_\Theta(C, N)$ "dual" to C. We have

$$\mathrm{Hom}^n_\Theta(C, N) = \mathrm{Hom}_\Theta(C_n, N) \tag{1.39}$$

as $\mathbb{S}$-graded vector spaces for each $n \in \mathbb{Z}$. The differential in the cochain complex is defined by $(\delta f)(c) = (-1)^{n+|f|} f(\partial c)$.

DEFINITION 1.75. Let Θ be an algebra graded on $\mathbb{S}$, and let C and D be chain complexes over the category of Θ-modules. We will define a chain complex $\overline{\mathrm{Hom}}^\Theta(C, D)$ over the category of $\mathbb{S}$-graded vector spaces. For each $n \in \mathbb{Z}$, the $\mathbb{S}$-graded vector space $\overline{\mathrm{Hom}}^\Theta_n(C, D)$ is defined by:

$$\overline{\mathrm{Hom}}^\Theta_n(C, D) = \prod_{p-q=n} \overline{\mathrm{Hom}}^\Theta(C_q, D_p). \tag{1.40}$$

The differential $\partial : \overline{\mathrm{Hom}}^\Theta_n(C, D) \to \overline{\mathrm{Hom}}^\Theta_{n-1}(C, D)$ is defined, for $f \in \overline{\mathrm{Hom}}^\Theta_n(C, D)$ and all $c \in C$, by: $(\partial f)(c) = \partial(f(c)) + (-1)^{n+|f|} f(\partial c)$.

Here too we will be using the special case that occurs when D is just a module N over Θ, regarded as a chain complex concentrated in homological degree 0. Then we get a chain complex $\overline{\mathrm{Hom}}^\Theta(C, N)$ "dual" to C, with

$$\overline{\mathrm{Hom}}^\Theta_{-n}(C, N) = \overline{\mathrm{Hom}}^\Theta(C_n, N) \tag{1.41}$$

and differential defined by $(\partial f)(c) = (-1)^{n+|f|} f(\partial c)$.

REMARK 1.76. If C, D are chain complexes of Θ-modules then $H_{*,*}\overline{\mathrm{Hom}}^\Theta(C, D)$ is the negative of the $(\mathbb{Z} \times \mathbb{S})$-graded vector space $H^{*,*}\,\mathrm{Hom}_\Theta(C, D)$.

REMARK 1.77. Suppose that Θ, Θ' are algebras; that C and C' are chain complexes over Θ and Θ', respectively; and that N and N' are modules over Θ and Θ', respectively. Then the definitions (1.23) extend in an obvious way to homomorphisms of cochain and chain complexes:

$$\nu : \mathrm{Hom}_\Theta(C, N) \otimes \mathrm{Hom}_{\Theta'}(C', N') \to \mathrm{Hom}_{\Theta\otimes\Theta'}(C \otimes C', N \otimes N') \tag{1.42a}$$

$$\overline{\nu} : \overline{\mathrm{Hom}}^\Theta(C, N) \otimes \overline{\mathrm{Hom}}^{\Theta'}(C', N') \to \overline{\mathrm{Hom}}^{\Theta\otimes\Theta'}(C \otimes C', N \otimes N'). \tag{1.42b}$$

REMARK 1.78. Suppose that Θ is a bialgebra. Suppose that C and C' are chain complexes over Θ, and that N, N' are right Θ-modules. Then the definitions (1.29) extend in an obvious way to homomorphisms of cochain and chain complexes:

$$\rho : \mathrm{Hom}_\Theta(C, N) \otimes \mathrm{Hom}_\Theta(C', N') \to \mathrm{Hom}_\Theta(C \otimes C', N \otimes N') \tag{1.43a}$$

$$\overline{\rho} : \overline{\mathrm{Hom}}^\Theta(C, N) \otimes \overline{\mathrm{Hom}}^\Theta(C', N') \to \overline{\mathrm{Hom}}^\Theta(C \otimes C', N \otimes N'). \tag{1.43b}$$

PROPOSITION 1.79. *Let C, D, G be chain complexes over the category of $\mathbb{S}$-graded vector spaces. Then there is an isomorphism of cochain complexes:*

$$\phi : \mathrm{Hom}(C \otimes D, G) \to \mathrm{Hom}(C, \overline{\mathrm{Hom}}(D, G)) \tag{1.44}$$

that is natural with respect to homomorphisms of the chain complexes C, D and G.

PROOF. The isomorphism in the category of $(\mathbb{Z} \times \mathbb{S})$-graded vector spaces is as described earlier in (1.4). But it is easy to check that the isomorphism commutes with the differentials as we have defined them above. □

REMARK 1.80. In Chapter 4 we will need a refinement of the above result. For each $p \in \mathbb{Z}$ one can make $C_p \otimes D$ into a chain complex by writing

$$\partial(c \otimes d) = (-1)^{p+|c|}\, c \otimes \partial d.$$

Corresponding to this chain complex is the dual cochain complex $\mathrm{Hom}(C_p \otimes D, G)$. Similarly fixing p one gets the cochain complex $\mathrm{Hom}(C_p, \overline{\mathrm{Hom}}(D, G))$. It is easy to see that the isomorphism of (1.44) restricts to an isomorphism of cochain complexes

$$\phi_p : \mathrm{Hom}(C_p \otimes D, G) \to \mathrm{Hom}(C_p, \overline{\mathrm{Hom}}(D, G)) \tag{1.45}$$

for each $p \in \mathbb{Z}$.

The conventions we have adopted here apply to the definitions of the derived functors of Hom. Thus, if Θ is an algebra graded on $\mathbb{S}$, and M a right Θ-module, we construct a projective resolution of M: $\cdots \to \mathcal{P}_n(M) \to \cdots \to \mathcal{P}_0(M) \to M \to 0$. We will write $\mathcal{P}(M)$ for the chain complex: $\cdots \to \mathcal{P}_n(M) \to \cdots \to \mathcal{P}_0(M)$. Then we have for every right Θ-module N the standard definition:

$$\mathrm{Ext}_\Theta^{*,*}(M, N) = H^{*,*}\,\mathrm{Hom}_\Theta(\mathcal{P}(M), N) \tag{1.46}$$

as vector spaces graded on $\mathbb{Z} \times \mathbb{S}$. We will also need the definition

$$\overline{\mathrm{Ext}}_{*,*}^{\Theta}(M, N) = H_{*,*}\overline{\mathrm{Hom}}^{\Theta}(\mathcal{P}(M), N), \tag{1.47}$$

where we are using (1.40), regarding N as a chain complex concentrated in homological degree zero. By Remark 1.76, $\overline{\mathrm{Ext}}_{*,*}^{\Theta}(M, N)$ is the negative of the $(\mathbb{Z} \times \mathbb{S})$-graded vector space $\mathrm{Ext}_\Theta^{*,*}(M, N)$. In particular, $\overline{\mathrm{Ext}}_{*,*}^{\Theta}(M, N)$ is concentrated in homological degrees less than or equal to zero.

8. Differential Algebras and Coalgebras

The material in this section is needed only for the treatment of the Eilenberg-Moore spectral sequence in Chapter 6. We include it here, since it draws on the material on chain complexes we have just discussed. In what follows the symbol l will refer not only to the ground field, but also to the chain complex that is isomorphic to l in degree zero, and is zero in all other homological degrees.

DEFINITION 1.81. By a differential algebra we mean a chain complex A over the category of ungraded vector spaces that is equipped with two chain maps: a unit $\eta : l \to A$ and a product $\mu : A \otimes A \to A$. These are required to preserve (homological) degree, and to give A the structure of a $\mathbb{Z}$-graded algebra.

DEFINITION 1.82. By a differential coalgebra we mean a chain complex B over the category of ungraded vector spaces, equipped with two chain maps: a counit $\epsilon : B \to l$ and a coproduct $\psi : B \to B \otimes B$. These are required to preserve (homological) degree, and to give B the structure of a $\mathbb{Z}$-graded coalgebra.

These definitions could easily be generalized to give notions of differential algebras and differential coalgebras over the category of $\mathbb{S}$-graded vector spaces, for any abelian group $\mathbb{S}$. But in this work we will need only the case $\mathbb{S} = 0$.

REMARK 1.83. Clearly if A is a differential algebra then its homology H_*A becomes a $\mathbb{Z}$-graded algebra, with unit $H\eta : l \to H_*A$ and product

$$H_*A \otimes H_*A \equiv H_*(A \otimes A) \xrightarrow{H_*\mu} H_*A.$$

If B is a differential coalgebra then H_*B becomes a $\mathbb{Z}$-graded coalgebra, with counit $H\epsilon : H_*B \to l$ and coproduct

$$H_*B \xrightarrow{H_*\psi} H_*(B \otimes B) \equiv H_*B \otimes H_*B.$$

DEFINITION 1.84. If A is a differential algebra, then by a differential right A-module we mean a chain complex C over the category of ungraded vector spaces, equipped with a chain map $\sigma : C \otimes A \to C$ that makes C into a right A-module. Differential left A-modules are defined similarly.

DEFINITION 1.85. Supposing A a differential algebra, C a differential right A-module and D a differential left A-module, then by $C \otimes_A D$ we mean the chain complex for which the underlying $\mathbb{Z}$-graded vector space is the usual tensor product of A-modules $C \otimes_A D$, with differential induced by that on $C \otimes D$. If C and D are both differential right A-modules, by $\operatorname{Hom}_A(C, D)$ we mean the sub-cochain complex of $\operatorname{Hom}(C, D)$ consisting of the homomorphisms of right A-modules, as these are defined in Definition 1.16.

REMARK 1.86. If A is a differential algebra and if C is a differential right A-module, then H_*C is a right H_*A-module under the action

$$H_*C \otimes H_*A \equiv H_*(C \otimes A) \xrightarrow{H_*\sigma} H_*C. \tag{1.48}$$

If C and D are both differential right A-modules there is an obvious mapping of graded vector spaces:

$$h : H^* \operatorname{Hom}_A(C, D) \to \operatorname{Hom}_{H_*A}(H_*C, H_*D). \tag{1.49}$$

REMARK 1.87. Suppose A a differential algebra and D a differential left A-module; then $\overline{\operatorname{Hom}}(D, l)$ becomes a differential right A-module under the rule $(fa)(d) = f(ad)$ for all $f \in \overline{\operatorname{Hom}}(D, l), a \in A, d \in D$. Notice the effect of our using $\overline{\operatorname{Hom}}(D, l)$ rather than $\operatorname{Hom}(D, l)$. With this choice we have got a chain, rather than a cochain complex; and the action $\overline{\operatorname{Hom}}(D, l) \otimes A \to \overline{\operatorname{Hom}}(D, l)$ is degree-preserving, as required by our definitions. Passing to homology we obtain a right action $\overline{\operatorname{Hom}}(H_*D, l) \otimes H_*A \to \overline{\operatorname{Hom}}(H_*D, l)$. This right action is obtained from the original left action $H_*A \otimes H_*D \to H_*D$ by the construction (1.25).

Now we have immediately:

PROPOSITION 1.88. *Let A be a differential algebra, C a differential right A-module, and D a differential left A-module. Then the isomorphism* (1.44) *restricts to an isomorphism of cochain complexes*

$$\phi : \mathrm{Hom}(C \otimes_A D, l) \to \mathrm{Hom}_A(C, \overline{\mathrm{Hom}}(D, l)) \tag{1.50}$$

that is natural with respect to mappings of the differential A-modules C and D.

We will use (1.49) and (1.50) in Chapter 6 to describe the E_2-term of the Eilenberg-Moore spectral sequence.

9. Simplicial Θ-Modules

The material in this section is central to the book. We review properties of the "shuffle" mapping of Eilenberg and Mac Lane, and of its homotopy inverse, the Alexander-Whitney map. We review properties of Steenrod's "cup-k products" which measure the deviation of the Alexander-Whitney map from commutativity. We single out a class of cup-k products that we call " special". These have filtration-preserving properties that will play a key role when, in the next chapter, we put Steenrod operations into spectral sequences.

In the course of discussing the Alexander-Whitney map we review the totally amazing Proposition 1.90, due to Eilenberg and Mac Lane. This is a technical result relating shuffle and Alexander-Whitney mappings, which we use in the next section to show that the Alexander-Whitney map is under certain circumstances a map of differential algebras. This material will be useful when we discuss the Eilenberg-Moore spectral sequence in Chapters 6 and 7.

Fix a bialgebra Θ over a field l, graded on the abelian group $\mathbb{S}$. Throughout this book we will be working with simplicial objects over the category of Θ-modules ("simplicial Θ-modules"). We understand by this that face and degeneracy operators preserve the $\mathbb{S}$-grading. If R is a simplicial Θ-module we will write CR for the associated chain complex over the category of Θ-modules: $C_nR = R_n$ with differential the alternating sum of the face operators.

If R and S are simplicial Θ-modules we write $R \times S$ for their cross product. This is the simplicial Θ-module defined by: $(R \times S)_n = R_n \otimes S_n$, with diagonal action of Θ, and face and degeneracy operators $d_i^{R\times S} = d_i^R \otimes d_i^S$ and $s_i^{R\times S} = s_i^R \otimes s_i^S$. The Θ-chain complexes $CR \otimes CS$ and $C(R \times S)$ are related by a pair of chain maps due to Eilenberg and Mac Lane:

$$CR \otimes CS \underset{f}{\overset{\nabla}{\rightleftarrows}} C(R \times S). \tag{1.51}$$

These are natural with respect to mappings of both the simplicial Θ-modules R and S. The "shuffle mapping" ∇ is given by equation (5.3) of [**19**, page 64]:

$$\nabla(r \otimes s) = \sum_{(\mu,\nu)} (-1)^{\epsilon(\mu)} s_{\nu_q} \cdots s_{\nu_1}(r) \otimes s_{\mu_p} \cdots s_{\mu_1}(s), \tag{1.52}$$

for $r \in R_p, s \in S_q$. The sum is taken over all (p, q) shuffles (μ, ν), and $\epsilon(\mu)$ is the integer $\sum_{1 \le i \le p} (\mu_i - (i-1))$. The mapping f generalizes from simplicial complexes to simplicial sets the Alexander-Whitney formula for cup-product at chain level. It is given by equation (2.9) of [**20**]:

$$f(r \otimes s) = \sum_{i=0}^{n} \tilde{F}^{n-i} r \otimes F^i s, \tag{1.53}$$

for all $r \in R_n, s \in S_n$. Here F is the "front face" operator defined by $F(s) = d_0 s$ for all $s \in S$; and $\tilde{F}$ is the "back face" operator defined by $\tilde{F}(r) = d_k r$ if $r \in R_k$. We will sometimes write $\bigtriangledown(R,S)$ and $f(R,S)$ for the mappings in (1.51). They were originally defined for the case in which R and S are simplicial modules over the ground field; but the naturality of $\bigtriangledown$ and f in that case implies that when R and S are Θ-modules, the mappings are linear over Θ as well. Both $\bigtriangledown$ and f are associative, in the obvious sense. $\bigtriangledown$ and f are chain-homotopy inverse to each other. The composition of isomorphisms

$$H_*CR \otimes H_*CS = H_*(CR \otimes CS) \xrightarrow{H_*\bigtriangledown} H_*C(R \times S) \tag{1.54}$$

is the usual identification of the tensor product of homologies with the homology of the cross product, although in the present context, (1.54) is an isomorphism of modules over Θ.

We record two further properties of $\bigtriangledown$ and f. These will be useful later in our discussion of the Eilenberg-Moore spectral sequence.

PROPOSITION 1.89. *Let R and S be simplicial Θ-coalgebras. Then the following diagram of chain complexes over Θ commutes:*

$$\begin{array}{ccc}
CR \otimes CS & \xrightarrow{\bigtriangledown(R,S)} & C(R \times S) \\
\Big\downarrow{\scriptstyle C\psi_R \otimes C\psi_S} & & \Big\downarrow{\scriptstyle C(\psi_R \times \psi_S)} \\
C(R \times R) \otimes C(S \times S) & \xrightarrow[\bigtriangledown(R \times R, S \times S)]{} & C(R \times R \times S \times S).
\end{array} \tag{1.55}$$

This is just a special case of the naturality of the shuffle mapping.

PROPOSITION 1.90. *Let R, R', S, S' be simplicial Θ-modules. Then the following diagram of chain complexes over Θ commutes:*

$$\begin{array}{ccc}
C(R \times S) \otimes C(R' \times S') & \xrightarrow{\bigtriangledown(R \times S, R' \times S')} & C(R \times S \times R' \times S') \\
\Big\downarrow{\scriptstyle f(R,S) \otimes f(R',S')} & & \Big\downarrow{\scriptstyle (1,3,2,4)} \\
CR \otimes CS \otimes CR' \otimes CS' & & C(R \times R' \times S \times S') \\
\Big\downarrow{\scriptstyle (1,3,2,4)} & & \Big\downarrow{\scriptstyle f(R \times R', S \times S')} \\
CR \otimes CR' \otimes CS \otimes CS' & \xrightarrow[\bigtriangledown(R,R') \otimes \bigtriangledown(S,S')]{} & C(R \times R') \otimes C(S \times S').
\end{array} \tag{1.56}$$

This highly non-obvious result is due to Eilenberg and Mac Lane. Its proof is contained in the proof of Theorem 3.2 of [**20**].

In the remainder of this section we take the ground field l to be $\mathbb{Z}/2$, and we review some of the material needed to define Steenrod operations.

We use the letter T to denote both the transposition mapping $T : R\times S \to S\times R$ of simplicial vector spaces, and the induced map $T : C(R\times S) \to C(S\times R)$ of chain complexes. We also use it for the transposition mapping $T : CR\otimes CS \to CS\otimes CR$.

DEFINITION 1.91. By a cup-k product we mean a family of vector space homomorphisms

$$D_k = D_k(R,S) : C_n(R\times S) \to (CR\otimes CS)_{n+k} \tag{1.57}$$

defined for all ordered pairs R, S of simplicial $\mathbb{S}$-graded vector spaces over $\mathbb{Z}/2$, natural in both variables with respect to homomorphisms of simplicial vector spaces, and defined for all non-negative values of the integers n, k. We require:

1. Each mapping (1.57) preserves "internal" degree (the degree indexed by $\mathbb{S}$).
2. D_0 is a chain map satisfying, for all $\sigma \in R_0, \tau \in S_0$

$$D_0(\sigma\otimes\tau) = \sigma\otimes\tau. \tag{1.58}$$

3. For each $k \geq 1$:

$$\partial D_k + D_k\partial = D_{k-1} + TD_{k-1}T. \tag{1.59}$$

The existence of a cup-k product, and its uniqueness up to a suitable notion of homotopy, is proved by Dold [**16**] using the method of acyclic models [**23**].

The next three propositions give some useful properties of cup-k products.

If V is a vector space over $\mathbb{Z}/2$, there is associated a "constant" simplicial vector space, which is isomorphic in each dimension to V, and for which the faces and degeneracies are all identity maps. We denote this simplicial vector space also by V.

PROPOSITION 1.92. *Let V, W be vector spaces, and R, S simplicial vector spaces. Write R' and S' for the simplicial vector spaces $R' = V\times R$ and $S' = W\times S$. Then if $\{D_k \mid k \geq 0\}$ is a cup-k product, the mapping $D_k(R', S') : C_n(R'\times S') \to (CR'\otimes CS')_{n+k}$ is given by:*

$$V\otimes R_n\otimes W\otimes S_n \xrightarrow{(1,3,2,4)} V\otimes W\otimes R_n\otimes S_n \xrightarrow{V\otimes W\otimes D_k(R,S)}$$

$$V\otimes W\otimes\left(\bigoplus_{i+j=n+k}(R_i\otimes S_j)\right) \xrightarrow{(1,3,2,4)} \bigoplus_{i+j=n+k}(V\otimes R_i\otimes W\otimes S_j).$$

This follows easily from the naturality of the mappings D_k.

PROPOSITION 1.93. *Let $\{D_k \mid k \geq 0\}$ be any cup-k product. Suppose Θ a bialgebra over $\mathbb{Z}/2$ and R and S simplicial Θ-modules. Then for all $n, k \geq 0$ the mapping $D_k : C_n(R\times S) \to (CR\otimes CS)_{n+k}$ is a homomorphism of Θ-modules.*

This also follows at once from the naturality of the D_k.

As a consequence we obtain for any Θ-module N and any simplicial Θ-modules R and S the $\mathbb{Z}/2$-homomorphisms dual to the D_k:

$$D_k^* : \mathrm{Hom}_\Theta^{n+k}(CR\otimes CS, N) \to \mathrm{Hom}_\Theta^n(C(R\times S), N), \tag{1.60}$$

where the dual complexes appearing here are defined as in (1.39). The homomorphisms (1.60) are degree-preserving homomorphisms of $\mathbb{S}$-graded vector spaces, and satisfy the duals of equations (1.59). We need to single out a class of cup-k products having special properties. Write Δ_n for the standard semi-simplicial n-simplex, and $\delta_n = [0, 1, \ldots, n] \in \Delta_n$ for the generating n-cell (see e.g., Chapter 1 of [**47**]). We will write $\Delta[n]$ for the simplicial vector space with basis Δ_n. Clearly the face operators of Δ_n carry non-degenerate simplices to non-degenerate simplices. We will write $\tilde{C}\Delta[n] \subseteq C\Delta[n]$ for the sub-chain complex spanned by the non-degenerate simplices. As is well-known, $\tilde{C}\Delta[n]$ is acyclic: $H_k\tilde{C}\Delta[n] = 0$ if $k > 0$, and the natural augmentation $\epsilon : H_0\tilde{C}\Delta[n] \to \mathbb{Z}/2$ is an isomorphism.

DEFINITION 1.94. A cup-k product $\{D_k \mid k \geq 0\}$ is called "special" if it satisfies all of the following:

1. The chain mapping $D_0 : C(R \times S) \to CR \otimes CS$ is the Alexander-Whitney mapping f of (1.53), for all simplicial vector spaces R, S.
2. The homomorphism $D_k : C_n(\Delta[n] \times \Delta[n]) \to (C\Delta[n] \otimes C\Delta[n])_{n+k}$ satisfies

(1.61) $$D_k(\delta_n \otimes \delta_n) \in \tilde{C}\Delta[n] \otimes \tilde{C}\Delta[n]$$

for all values of k, n; and

(1.62) $$D_n(\delta_n \otimes \delta_n) = \delta_n \otimes \delta_n.$$

PROPOSITION 1.95. *There exists a special cup-k product.*

The proof is by the method of acyclic models [**23**], and is essentially that given by Dold [**16**]. Another good exposition can be found in [**82**]: one need only replace Spanier's models Δ_n by the models $\Delta_n \times \Delta_n$, and his standard simplex $\delta_n \in C\Delta[n]$ by $\delta_n \otimes \delta_n \in C(\Delta[n] \times \Delta[n])$. One uses the acyclicity of $\tilde{C}\Delta[n]$.

PROPOSITION 1.96. *Suppose $\{D_k \mid k \geq 0\}$ a special cup-k product. Then:*

1. *For all simplicial vector spaces R, S, and all integers $k, n \geq 0$, the image of the mapping $D_k : C_n(R \times S) \to (CR \otimes CS)_{n+k}$ is contained in the subspace $\bigoplus_{i,j \leq n}(C_iR \otimes C_jS)$. In particular $D_k(C_n(R \times S)) = 0$ if $k > n$.*
2. *For each $n \geq 0$ the map $D_n : C_n(R \times S) \to (CR \otimes CS)_{2n}$ is given by $D_n(\sigma \otimes \tau) = \sigma \otimes \tau$ for all $\sigma \in R_n$, $\tau \in S_n$.*

PROOF. To prove the first statement, suppose given $\sigma \in R_n$, $\tau \in S_n$. We write $f_\sigma : \Delta[n] \to R$ and $f_\tau : \Delta[n] \to S$ for the unique maps of simplicial vector spaces satisfying $f_\sigma(\delta_n) = \sigma$ and $f_\tau(\delta_n) = \tau$. Then by the naturality of the cup-k product we have a commutative diagram of vector spaces:

$$\begin{array}{ccc}
C_n(\Delta[n] \times \Delta[n]) & \xrightarrow{D_k(\Delta[n], \Delta[n])} & (C\Delta[n] \otimes C\Delta[n])_{n+k} \\
\Big\downarrow {\scriptstyle C(f_\sigma \times f_\tau)} & & \Big\downarrow {\scriptstyle Cf_\sigma \otimes Cf_\tau} \\
C_n(R \times S) & \xrightarrow[D_k(R,S)]{} & (CR \otimes CS)_{n+k}.
\end{array}$$

Applying this diagram to $\delta_n \otimes \delta_n$ in the upper left we find:

(1.63) $$D_k(R, S)(\sigma \otimes \tau) = (Cf_\sigma \otimes Cf_\tau)D_k(\Delta[n], \Delta[n])(\delta_n \otimes \delta_n).$$

But we have from (1.61) that $D_k(\Delta[n], \Delta[n])(\delta_n \otimes \delta_n) \in \tilde{C}\Delta[n] \otimes \tilde{C}\Delta[n]$; and the chain complex $\tilde{C}\Delta[n]$ is zero in degrees larger than n. So (1.63) imples the first statement of our proposition. Similarly, the naturality of the cup-k product together with (1.62) implies the second statement. □

10. Homology of Simplicial Sets and Simplicial Groups

The material in this section is needed only for our treatment of the Eilenberg-Moore spectral sequence, in Chapters 6 and 7. The chains and cochains on simplicial sets and simplicial groups inherit certain products and coproducts from the shuffle and Alexander-Whitney mappings. Similarly, the chains on a simplicial set that has an action of a simplicial group $\mathcal{G}$ inherit an action of $C_*\mathcal{G}$. We discuss these structures in this section. None of these results are new, and some are well known. But we call the reader's attention particularly to Propositions 1.97, 1.100, 1.101, and 1.103, all of which are consequences of the remarkable property (1.56) of the Alexander-Whitney map.

If $\mathcal{R}$ is a simplicial set we will write R for the simplicial vector space over the ground field l of which $\mathcal{R}$ is the basis. Then the usual homology and cohomology of $\mathcal{R}$ with coefficients in the ground field l are:

$$H_*\mathcal{R} = H_*(CR) \tag{1.64}$$

$$H^*\mathcal{R} = H^* \operatorname{Hom}(CR, l). \tag{1.65}$$

We write $j_{\mathcal{R}}$ for the standard inclusion of a vector space into its double dual:

$$j_{\mathcal{R}} : H_*\mathcal{R} \to \operatorname{Hom}(H^*\mathcal{R}, l). \tag{1.66}$$

We will also write $K_*\mathcal{R}$ for the negatively graded cohomology of $\mathcal{R}$:

$$K_*\mathcal{R} = H_*\overline{\operatorname{Hom}}(CR, l). \tag{1.67}$$

Let $\mathcal{R}$ and $\mathcal{S}$ be simplicial sets, and R and S the simplicial vector spaces they span. Then the simplicial vector space spanned by $\mathcal{R} \times \mathcal{S}$ is the cross product $R \times S$ as it has been defined at the beginning of Section 9. So in this case the composition (1.54) is giving the standard isomorphism:

$$H_*\mathcal{R} \otimes H_*\mathcal{S} = H_*(CR \otimes CS) \xrightarrow{H_*\nabla} H_*C(R \times S) = H_*(\mathcal{R} \times \mathcal{S}). \tag{1.68}$$

We record also its inverse:

$$H_*(\mathcal{R} \times \mathcal{S}) = H_*C(R \times S) \xrightarrow{H_*f} H_*(CR \otimes CS) = H_*\mathcal{R} \otimes H_*\mathcal{S}. \tag{1.69}$$

Here ∇ and f are as in (1.51), (1.53).

The construction dual to (1.69) is the cross product in cohomology. One defines the cross product by forming a composition of cochain maps:

$$\operatorname{Hom}(CR, l) \otimes \operatorname{Hom}(CS, l) \xrightarrow{\nu} \operatorname{Hom}(CR \otimes CS, l) \xrightarrow{f^*} \operatorname{Hom}(C(R \times S), l). \tag{1.70}$$

with ν as in (1.42a). The induced map in cohomology is the cross product:

$$H^*\mathcal{R} \otimes H^*\mathcal{S} \xrightarrow{cr} H^*(\mathcal{R} \times \mathcal{S}). \tag{1.71}$$

The mappings (1.69) and (1.71) are related by the commutative diagram:

$$
\begin{array}{ccc}
H_*(\mathcal{R}\times\mathcal{S}) & \xrightarrow{H_*f} & H_*\mathcal{R}\otimes H_*\mathcal{S} \\
 & & \Big\downarrow j_{\mathcal{R}}\otimes j_{\mathcal{S}} \\
\Big\downarrow j_{\mathcal{R}\times\mathcal{S}} & & \mathrm{Hom}(H^*\mathcal{R},l)\otimes\mathrm{Hom}(H^*\mathcal{S},l) \\
 & & \Big\downarrow \nu \\
\mathrm{Hom}(H^*(\mathcal{R}\times\mathcal{S}),l) & \xrightarrow[cr^*]{} & \mathrm{Hom}(H^*\mathcal{R}\otimes H^*\mathcal{S},l).
\end{array}
\tag{1.72}
$$

If $\mathcal{R}$ is a simplicial set, the associated simplicial vector space becomes a simplicial coalgebra, with coproduct and counit defined by

$$\psi(r) = r\otimes r \tag{1.73a}$$

$$\epsilon(r) = 1 \tag{1.73b}$$

for all $r\in\mathcal{R}$. It is also known classically that one can use these constructions to make CR into a differential coalgebra, as follows. Clearly the homomorphisms $\psi : R_n \to R_n\otimes R_n$ can be viewed as defining a mapping of simplicial vector spaces $\psi_R : R\to R\times R$. Then the composition:

$$CR \xrightarrow{C\psi_R} C(R\times R) \xrightarrow{f} CR\otimes CR \tag{1.74}$$

is a chain map defining a natural, coassociative coproduct on CR. One defines the augmentation $\epsilon : CR\to l$ by setting $\epsilon(r) = 1$ for each $r\in\mathcal{R}_0$.

We will need also a dual construction. If $\mathcal{F}$ is a simplicial set then $\overline{\mathrm{Hom}}(CF,l)$ becomes a differential algebra, with product given by the composition:

$$\overline{\mathrm{Hom}}(CF,l)\otimes\overline{\mathrm{Hom}}(CF,l) \xrightarrow{\overline{\nu}} \overline{\mathrm{Hom}}(CF\otimes CF,l) \xrightarrow{f^*} \overline{\mathrm{Hom}}(C(F\times F),l) \text{---} \xrightarrow{(C\psi_F)^*} \overline{\mathrm{Hom}}(CF,l). \tag{1.75}$$

As reviewed in Section 8, the homology of a differential coalgebra is itself a $\mathbb{Z}$-graded coalgebra. So for each simplicial set $\mathcal{R}$, $H_*\mathcal{R}$ is a $\mathbb{Z}$-graded coalgebra. The natural coproduct:

$$\psi_{\mathcal{R}} : H_*\mathcal{R}\to H_*\mathcal{R}\otimes H_*\mathcal{R}, \tag{1.76}$$

is the mapping in homology induced by (1.74); and the augmentation $\epsilon_{\mathcal{R}} : H_*\mathcal{R}\to l$ is induced by $\epsilon : CR\to l$. The coproduct can also be regarded as induced by the diagonal. $\psi_{\mathcal{R}}$ is the composition:

$$H_*\mathcal{R} \xrightarrow{H_*d} H_*(\mathcal{R}\times\mathcal{R}) \xrightarrow{H_*f} H_*\mathcal{R}\otimes H_*\mathcal{R}. \tag{1.77}$$

Similarly the mapping in homology induced by (1.75) gives a natural product:

$$\mu_{\mathcal{F}} : K_*\mathcal{F}\otimes K_*\mathcal{F}\to K_*\mathcal{F}. \tag{1.78}$$

Suppose now that $\mathcal{G}$ is a simplicial group, with product $\mu : \mathcal{G} \times \mathcal{G} \to \mathcal{G}$. Then μ defines a mapping of the associated simplicial vector spaces: $\mu_G : G \times G \to G$. We make CG into a differential algebra in the classical way. The product is the composition:

$$CG \otimes CG \xrightarrow{\nabla} C(G \times G) \xrightarrow{C\mu_G} CG \tag{1.79}$$

with ∇ the shuffle mapping of (1.51). The unit $\eta : l \to CG$ sends 1 to the unit element of $\mathcal{G}_0$. As reviewed in Section 8 the homology of a differential algebra is a $\mathbb{Z}$-graded algebra. So for each simplicial group $\mathcal{G}$, $H_*\mathcal{G}$ is a $\mathbb{Z}$-graded algebra. The product:

$$\mu_{\mathcal{G}} : H_*\mathcal{G} \otimes H_*\mathcal{G} \to H_*\mathcal{G}, \tag{1.80}$$

is the mapping in homology induced by (1.79); and the unit $\eta_{\mathcal{G}} : l \to H_*\mathcal{G}$ is induced by $\eta : l \to CG$.

PROPOSITION 1.97. *Let $\mathcal{G}$ and $\mathcal{G}'$ be simplicial groups. Then the Alexander-Whitney mapping $f : C(G \times G') \to CG \otimes CG'$ of (1.53) is a homomorphism of differential algebras.*

This is [**20**], Theorem 3.2. It follows at once from Proposition 1.90 and the naturality of f.

If $\mathcal{G}$ is a simplicial group then clearly the diagonal $d : \mathcal{G} \to \mathcal{G} \times \mathcal{G}$ is a mapping of simplicial groups, so $C\psi_G : CG \to C(G \times G)$ is a homomorphism of differential algebras. So we have from Proposition 1.97 that the following composition:

$$CG \xrightarrow{C\psi_G} C(G \times G) \xrightarrow{f} CG \otimes CG \tag{1.81}$$

is also a homomorphism of differential algebras. But this is (1.74) with $R = G$: it is the composition one uses to define the coproduct $\psi_{\mathcal{G}} : H_*\mathcal{G} \to H_*\mathcal{G} \otimes H_*\mathcal{G}$. We conclude that $\psi_{\mathcal{G}}$ is an algebra homomorphism, and so, as is classical, $H_*\mathcal{G}$ becomes a bialgebra. Finally we write $\nu : \mathcal{G} \to \mathcal{G}$ for the mapping defined by $\nu(g) = g^{-1}$ for all $g \in \mathcal{G}$. This mapping of simplicial sets defines a linear mapping of vector spaces $\nu_{\mathcal{G}} : H_*\mathcal{G} \to H_*\mathcal{G}$ which satisfies the axiom (1.26) for an antipode. So we have the classical result that $H_*\mathcal{G}$ is a Hopf algebra with structure maps $\{\mu_{\mathcal{G}}, \eta_{\mathcal{G}}, \psi_{\mathcal{G}}, \epsilon_{\mathcal{G}}, \nu_{\mathcal{G}}\}$. Suppose now that $\mathcal{G}$ is a simplicial group and $\mathcal{E}$ a simplicial set on which $\mathcal{G}$ acts from the right (a "right $\mathcal{G}$-space"), with action $\sigma : \mathcal{E} \times \mathcal{G} \to \mathcal{E}$. Then σ defines also a mapping of the associated simplicial vector spaces $\sigma_E : E \times G \to E$, and CE becomes a differential right module over the differential algebra CG, with action given by the composition of chain maps:

$$CE \otimes CG \xrightarrow{\nabla} C(E \times G) \xrightarrow{C\sigma_E} CE. \tag{1.82}$$

The induced map in homology as reviewed in Remark 1.86 is in this case

$$\sigma_{\mathcal{E}} : H_*\mathcal{E} \otimes H_*\mathcal{G} \to H_*\mathcal{E}. \tag{1.83}$$

This gives $H_*\mathcal{E}$ the structure of a right module over $H_*\mathcal{G}$.

Similarly if $\mathcal{F}$ is a left $\mathcal{G}$-space then CF becomes a differential left module over CG. It follows from Remark 1.87 that the dual chain complex $\overline{\mathrm{Hom}}(CF, l)$ becomes a differential right module over CG. The action

$$\overline{\mathrm{Hom}}(CF, l) \otimes CG \to \overline{\mathrm{Hom}}(CF, l) \tag{1.84}$$

on passage to homology gives an action:

$$K_*\mathcal{F} \otimes H_*\mathcal{G} \to K_*\mathcal{F}. \tag{1.85}$$

Clearly the diagonal $d : \mathcal{E} \to \mathcal{E} \times \mathcal{E}$ is a mapping of right $\mathcal{G}$-spaces, where $\mathcal{G}$ acts on $\mathcal{E} \times \mathcal{E}$ through the diagonal $d : \mathcal{G} \to \mathcal{G} \times \mathcal{G}$. As a consequence of this, and of the naturality of the shuffle, we have:

PROPOSITION 1.98. *Let $\mathcal{G}$ be a simplicial group, and $\mathcal{E}$ a right $\mathcal{G}$-space. Then $C\psi_E : CE \to C(E \times E)$ is a homomorphism of differential right CG-modules, where CG acts on $C(E \times E)$ through the homomorphism of differential algebras $C\psi_G : CG \to C(G \times G)$.*

Suppose now $\mathcal{G}$ a simplicial group, $\mathcal{E}$ a right $\mathcal{G}$-space and $\mathcal{F}$ a left $\mathcal{G}$-space. One consequence of Proposition 1.98 is that the tensor product of chain mappings

$$C\psi_E \otimes C\psi_F : CE \otimes CF \to C(E \times E) \otimes C(F \times F) \tag{1.86}$$

passes to a quotient chain mapping:

$$C\psi_E \otimes_{CG} C\psi_F : CE \otimes_{CG} CF \to C(E \times E) \otimes_{C(G\times G)} C(F \times F). \tag{1.87}$$

We observe also that, as a consequence of the associativity of the shuffle, the chain map $\nabla(E, F) : CE \otimes CF \to C(E \times F)$ passes to a quotient

$$\nabla = \nabla(E, F) : CE \otimes_{CG} CF \to C(E \times_G F). \tag{1.88}$$

Here we are writing $E \times_G F$ for the simplicial vector space with basis $\mathcal{E} \times_{\mathcal{G}} \mathcal{F}$. The quotient chain maps that we have just defined are related by:

PROPOSITION 1.99. *Let $\mathcal{G}$ be a simplicial group, $\mathcal{E}$ a right $\mathcal{G}$-space and $\mathcal{F}$ a left $\mathcal{G}$-space. Then the following diagram of chain mappings commutes:*

(1.89)

$$\begin{array}{ccc}
CE \otimes_{CG} CF & \xrightarrow{\nabla(E,F)} & C(E \times_G F) \\
\downarrow{\scriptstyle C\psi_E \otimes_{CG} C\psi_F} & & \downarrow{\scriptstyle C(\psi_E \times_G \psi_F)} \\
C(E \times E) \otimes_{C(G\times G)} C(F \times F) & \xrightarrow[\nabla(E \times E, F \times F)]{} & C((E \times E) \times_{(G\times G)} (F \times F)).
\end{array}$$

PROOF. The commutativity of this diagram follows from that of (1.55) after passing to quotients. □

We have just seen some relationships between simplicial $\mathcal{G}$-actions and the shuffle. Now we develop some relationships between $\mathcal{G}$-actions and the Alexander-Whitney map. We have already seen (Proposition 1.97) that if $\mathcal{G}$ and $\mathcal{G}'$ are simplicial groups then $f(G, G') : C(G \times G') \to CG \otimes CG'$ is a homomorphism of differential algebras. Now we suppose we are also given a right $\mathcal{G}$-space $\mathcal{E}$, and a right $\mathcal{G}'$-space $\mathcal{E}'$.

PROPOSITION 1.100. *The Alexander-Whitney mapping*

$$f(E,E'): C(E\times E') \to CE\otimes CE'$$

is a mapping of differential right modules over the differential algebra $C(G\times G')$, *where* $C(G\times G')$ *is acting on* $CE\otimes CE'$ *through the homomorphism of differential algebras* $f(G,G'): C(G\times G') \to CG\otimes CG'$.

We prove this just as we proved Proposition 1.97. It follows easily from Proposition 1.90 and the naturality of f.

As a consequence of Propositions 1.98 and 1.100 we have:

PROPOSITION 1.101. *Let* $\mathcal{G}$ *be a simplicial group, and* $\mathcal{E}$ *a right* $\mathcal{G}$*-space. Then the composition* (1.74) *of chain mappings*

$$CE \xrightarrow{C\psi_E} C(E\times E) \xrightarrow{f(E,E)} CE\otimes CE \tag{1.90}$$

that defines CE *as a differential coalgebra is a homomorphism of right* CG *modules, where* CG *is acting on* $CE\otimes CE$ *through the homomorphism of differential algebras*

$$CG \xrightarrow{C\psi_G} C(G\times G) \xrightarrow{f(G,G)} CG\otimes CG. \tag{1.91}$$

As a consequence of Proposition 1.101 we have:

PROPOSITION 1.102. *Let* $\mathcal{G}$ *be a simplicial group, and* $\mathcal{E}$ *a right* $\mathcal{G}$*-space. Then with the coproduct* $\psi_{\mathcal{E}}: H_*\mathcal{E} \to H_*\mathcal{E}\otimes H_*\mathcal{E}$ *as it is defined in* (1.76), *and action* $\sigma_{\mathcal{E}}: H_*\mathcal{E}\otimes H_*\mathcal{G} \to H_*\mathcal{E}$ *as defined in* (1.83), $H_*\mathcal{E}$ *becomes a right* $H_*\mathcal{G}$*-coalgebra.*

We will need duals of Propositions 1.101 and 1.102. We have already observed that if $\mathcal{F}$ is a left $\mathcal{G}$-space then $\overline{\mathrm{Hom}}(CF,l)$ becomes a differential right module over CG, as in (1.84).

PROPOSITION 1.103. *Suppose* $\mathcal{G}$ *a simplicial group and* $\mathcal{F}$ *a left* $\mathcal{G}$*-space. Then the composition* (1.75) *of chain mappings that defines* $\overline{\mathrm{Hom}}(CF,l)$ *as a differential algebra is a homomorphism of right* CG*-modules, where* CG *is acting on* $\overline{\mathrm{Hom}}(CF,l)\otimes\overline{\mathrm{Hom}}(CF,l)$ *through the homomorphism* (1.91) *of differential algebras.*

PROOF. It is easy to check that $\overline{\nu}$ of (1.75) is a homomorphism of right modules over $CG\otimes CG$. Therefore it is also a homomorphism of right modules over CG, where CG is acting on both $\overline{\mathrm{Hom}}(CF,l)\otimes\overline{\mathrm{Hom}}(CF,l)$ and $\overline{\mathrm{Hom}}(CF\otimes CF,l)$ through (1.91). That the composition $(C\psi_F)^*\cdot f^*$ of (1.75) is also linear over CG is just the dual of Proposition 1.101. □

As a corollary of Proposition 1.103 we have:

COROLLARY 1.104. *Suppose* $\mathcal{G}$ *a simplicial group and* $\mathcal{F}$ *a left* $\mathcal{G}$*-space. With the action of* $H_*\mathcal{G}$ *upon* $K_*\mathcal{F}$ *as defined by* (1.85), $K_*\mathcal{F}$ *becomes a right* $H_*\mathcal{G}$*-algebra.*

We remark that it was not necessary to go to "the chain level" to prove Corollary 1.104. It could have been proved directly within the context of homology, by dualizing Proposition 1.102, and using Proposition 1.31. But the chain level Proposition 1.103 will be useful in its own right (see Proposition 2.27 below).

Of course Proposition 1.100 remains true if the simplicial sets $\mathcal{E}$ and $\mathcal{E}'$ on which $\mathcal{G}$ and $\mathcal{G}'$ act from the right, are replaced by simplicial sets $\mathcal{F}$, $\mathcal{F}'$ on which

$\mathcal{G}$ and $\mathcal{G}'$ act from the left. It follows that the tensor product of chain mappings: $f(E,E')\otimes f(F,F') : C(E\times E')\otimes C(F\times F') \to (CE\otimes CE')\otimes(CF\otimes CF')$ passes to a well-defined quotient:

(1.92)
$$C(E\times E')\otimes_{C(G\times G')} C(F\times F') \xrightarrow{f(E,E')\otimes f(F,F')} (CE\otimes CE')\otimes_{(CG\otimes CG')}(CF\otimes CF').$$

Recall that the shuffle also passes to a quotient mapping, as in (1.88), so that all the chain maps appearing in the following proposition are well-defined.

PROPOSITION 1.105. *Let $\mathcal{G}$ and $\mathcal{G}'$ be simplicial groups. Suppose given $\mathcal{E}$ and $\mathcal{E}'$, right spaces over $\mathcal{G}$ and $\mathcal{G}'$ respectively; and suppose given $\mathcal{F}$ and $\mathcal{F}'$, left spaces over $\mathcal{G}$ and $\mathcal{G}'$ respectively. Then the following diagram of chain mappings commutes:*

(1.93)
$$\begin{array}{ccc}
C(E\times E')\otimes_{C(G\times G')} C(F\times F') & \xrightarrow{\nabla(E\times E',F\times F')} & C((E\times E')\times_{(G\times G')}(F\times F')) \\
\Big\downarrow{\scriptstyle f(E,E')\otimes f(F,F')} & & \Big\downarrow{\scriptstyle (1,3,2,4)} \\
(CE\otimes CE')\otimes_{(CG\otimes CG')}(CF\otimes CF') & & C((E\times_G F)\times(E'\times_{G'} F')) \\
\Big\downarrow{\scriptstyle (1,3,2,4)} & & \Big\downarrow{\scriptstyle f(E\times_G F,E'\times_{G'} F')} \\
(CE\otimes_{CG} CF)\otimes(CE'\otimes_{CG'} CF') & \xrightarrow[\nabla(E,F)\otimes\nabla(E',F')]{} & C(E\times_G F)\otimes C(E'\times_{G'} F').
\end{array}$$

This follows from (1.56) by passage to quotients.

We conclude this section with a study of the cohomology of spaces of the form $\mathcal{E}\times_{\mathcal{G}}\mathcal{F}$.

Suppose $\mathcal{G}$ a simplicial group, $\mathcal{E}$ is a right $\mathcal{G}$-space and $\mathcal{F}$ a left $\mathcal{G}$-space. We begin by taking duals in (1.88) and then passing to cohomology. We obtain natural homomorphisms

(1.94) $$\nabla^* = \nabla^*(E,F) : \operatorname{Hom}(C(E\times_G F),l) \to \operatorname{Hom}(CE\otimes_{CG} CF,l)$$

and

(1.95) $$H^*(\nabla^*) : H^*(\mathcal{E}\times_{\mathcal{G}}\mathcal{F}) \to H^*\operatorname{Hom}(CE\otimes_{CG} CF,l).$$

We apply to the right-hand side of (1.95) the formalism of Section 8. As in Remark 1.87 the chain complex $\overline{\operatorname{Hom}}(CF,l)$ is a differential right module over CG. So $K_*\mathcal{F} = H_*\overline{\operatorname{Hom}}(CF,l)$, as defined in (1.67), becomes a right module over $H_*\mathcal{G}$. Now we have from (1.49) and (1.50) the natural homomorphisms:

(1.96) $$H^*\operatorname{Hom}(CE\otimes_{CG} CF,l) \xrightarrow{H^*(\phi)} H^*(\operatorname{Hom}_{CG}(CE,\overline{\operatorname{Hom}}(CF,l)))\text{---}$$
$$\xrightarrow{h} \operatorname{Hom}_{H_*\mathcal{G}}(H_*\mathcal{E},K_*\mathcal{F}).$$

The composition in (1.96) can be further composed with $H^*(\nabla^*)$ of (1.95) to give a homomorphism of vector spaces:

$$hH^*(\phi)H^*(\nabla^*) : H^*(\mathcal{E} \times_{\mathcal{G}} \mathcal{F}) \to \operatorname{Hom}_{H_*\mathcal{G}}(H_*\mathcal{E}, K_*\mathcal{F}) \tag{1.97}$$

natural with respect to mappings of the simplicial $\mathcal{G}$-spaces $\mathcal{E}, \mathcal{F}$.

We will consider circumstances under which (1.97) is an isomorphism.

DEFINITION 1.106. Let $\mathcal{G}$ be a simplicial group. By an extended right $\mathcal{G}$-space. we mean a $\mathcal{G}$-space $\mathcal{E}$ of the form $\mathcal{E} = \mathcal{R} \times \mathcal{G}$, where $\mathcal{R}$ is any simplicial set, and the $\mathcal{G}$-action on $\mathcal{E}$ is given by $(r, g)g' = (r, gg')$.

LEMMA 1.107. *Let $\mathcal{E}$ be an extended right $\mathcal{G}$-space and $\mathcal{F}$ any left $\mathcal{G}$-space. Then the mapping* (1.97) *is an isomorphism of graded vector spaces.*

PROOF. Suppose $\mathcal{E} = \mathcal{R} \times \mathcal{G}$, and let $j : \mathcal{R} \to \mathcal{E}$ be the mapping of simplicial sets defined in each dimension p by $j(r) = (r, e_p)$ for each $r \in \mathcal{R}_p$, where e_p is the identity element of the group $\mathcal{G}_p$. Then j defines in an obvious way chain maps: $C(j \times F) : C(R \times F) \to C(E \times_G F)$; $Cj : CR \to CE$; $Cj \otimes CF : CR \otimes CF \to CE \otimes_{CG} CF$; and each of the squares in the following diagram commutes:

$$\begin{array}{ccc}
H^*\operatorname{Hom}(C(E \times_G F), l) & \xrightarrow{\;H^*C(j \times F)\;} & H^*\operatorname{Hom}(C(R \times F), l) \\
\Big\downarrow{\scriptstyle H^*(\nabla^*)} & & \Big\downarrow{\scriptstyle H^*(\nabla^*)} \\
H^*\operatorname{Hom}(CE \otimes_{CG} CF, l) & \xrightarrow{\;H^*(Cj \otimes CF)\;} & H^*\operatorname{Hom}(CR \otimes CF, l) \\
\Big\downarrow{\scriptstyle H^*(\phi)} & & \Big\downarrow{\scriptstyle H^*(\phi)} \\
H^*\operatorname{Hom}_{CG}(CE, \overline{\operatorname{Hom}}(CF, l)) & \xrightarrow{\;H^*(Cj)\;} & H^*\operatorname{Hom}(CR, \overline{\operatorname{Hom}}(CF, l)) \\
\Big\downarrow{\scriptstyle h} & & \Big\downarrow{\scriptstyle h} \\
\operatorname{Hom}_{H_*\mathcal{G}}(H_*\mathcal{E}, K_*\mathcal{F}) & \xrightarrow{\;(Hj)^*\;} & \operatorname{Hom}(H_*\mathcal{R}, K_*\mathcal{F}).
\end{array}$$

Here the vertical composition $hH^*(\phi)H^*(\nabla^*)$ on the right hand side of the diagram is just the well-known isomorphism $H^*(\mathcal{R} \times \mathcal{F}) = \operatorname{Hom}(H_*\mathcal{R} \otimes H_*\mathcal{F}, l) = \operatorname{Hom}(H_*R, K_*\mathcal{F})$. Also the horizontal arrows $H^*C(j \times F)$ and $(Hj)^*$ are isomorphisms, so the lemma follows. □

11. Cotriples, Simplicial Objects, and Projective Resolutions

If $\mathcal{C}$ is a category we will consider cotriples $\{T, d, s\}$ on $\mathcal{C}$ in the sense of Eilenberg and Moore [**22**]. Thus, $T : \mathcal{C} \to \mathcal{C}$ is a functor, $d : T \to Id$ is a natural transformation from T to the identity that is called the counit; and $s : T \to TT$ is a natural transformation called the coproduct. These are required to satisfy the left and right counitary laws, and the coassociative law, as in [**22**].

It is well-known that associated to any cotriple $\{T, d, s\}$ on a category $\mathcal{C}$ there is a functor W from $\mathcal{C}$ to the category of simplicial objects over $\mathcal{C}$. In fact, if Y

is an object of $\mathcal{C}$ one defines $W(Y)$ by setting $W_n(Y) = T^{n+1}(Y)$; face operators being given by the formulae: $d_i = T^i d T^{n-i} : T^{n+1}(Y) \to T^n(Y)$ and degeneracy operators by: $s_i = T^i s T^{n-i} : T^{n+1}(Y) \to T^{n+2}(Y)$.

If W is any simplicial object over a category $\mathcal{C}$, an augmentation of W is a mapping $\lambda : W_0 \to Y$ in the category satisfying $\lambda d_1 = \lambda d_0 : W_1 \to Y$. We will sometimes write $Y = W_{-1}$.

If $\{T, d, s\}$ is a cotriple defined on a category $\mathcal{C}$, and if Y is an object of $\mathcal{C}$, then the simplicial object $W(Y)$ comes equipped with a natural augmentation: $\lambda = d : (W_0(Y) = T(Y)) \to Y$. We give three examples of the construction W and associated augmentation.

EXAMPLE 1.108. Suppose Ξ a Hopf algebra, and M a right Ξ-module. Define a Ξ-module $T(M)$ by setting $T(M) = M \otimes \Xi$: the tensor product of Ξ-modules as described above. The mapping $d : M \otimes \Xi \to M$ given by $d = M \otimes \epsilon$ is a morphism of Ξ-modules, so d is a natural transformation of functors $d : T \to Id$. Also the mapping $s : M \otimes \Xi \to M \otimes \Xi \otimes \Xi$ given by $s = M \otimes \psi$ is a natural transformation $s : T \to TT$. Then $\{T, d, s\}$ is a cotriple. Associated to this cotriple is a functor W^Ξ from the category of Ξ-modules to the category of simplicial Ξ-modules. Associated to the simplicial module $W^\Xi(M)$ is the chain complex $CW^\Xi(M)$; and the augmentation $\lambda : W_0^\Xi(M) \to M$ described above defines an augmentation $\lambda : C_0(W^\Xi M) \to M$ of the chain complex $CW^\Xi(M)$. It is well-known that the augmented chain complex $CW^\Xi(M)$ is acyclic, but we record the argument here because we will later have to generalize beyond its classical context. Since $W_{p-1}^\Xi(M) = M \otimes \Xi^{\otimes p}$ for each $p \geq 0$, we can define a homomorphism $h : W_{p-1}^\Xi(M) \to W_p^\Xi(M)$ of vector spaces over the ground field l by setting:

$$h(m \otimes \xi_0 \otimes \cdots \otimes \xi_{p-1}) = m \otimes \xi_0 \otimes \cdots \otimes \xi_{p-1} \otimes 1. \tag{1.98}$$

Clearly for all $p \geq 0$, $d_0 h$ is the identity map on $W_{p-1}^\Xi(M)$, and $d_{i+1} h = h d_i$ as mappings from $W_p^\Xi(M)$ to itself for all i, $0 \leq i \leq p$. So if we regard the functions h as homomorphisms from $C_{p-1} W^\Xi(M)$ to $C_p W^\Xi(M)$ they define a contracting homotopy for the augmented chain complex $CW^\Xi(M) \to M$. It now follows from Proposition 1.24 that $CW^\Xi(M)$ is a free resolution of M as a Ξ-module. $CW^\Xi(M)$ is a version of the acyclic bar construction of Eilenberg and Mac Lane [**19, 42**].

EXAMPLE 1.109. Suppose Ξ a Hopf algebra and M a right Ξ-coalgebra. Then we can write $TM = M \otimes \Xi$ for the tensor product of Ξ-coalgebras as described in Proposition 1.29. T is then a functor from the category of Ξ-coalgebras to itself, and the mappings $d : TM \to M$, $s : TM \to TTM$, defined respectively by $d = M \otimes \epsilon$, $s = M \otimes \psi$ are natural transformations $d : T \to Id$ and $s : T \to TT$. The assumption that Ξ is cocommutative assures that s is a morphism of coalgebras. Then $\{T, d, s\}$ is a cotriple on the category of Ξ-coalgebras. The associated functor W^Ξ carries any Ξ-coalgebra M to a simplicial Ξ-coalgebra $W^\Xi(M)$ augmenting to M. The contracting homotopies $h : W_{p-1}^\Xi(M) \to W_p^\Xi(M)$ defined by (1.98) are in this context homomorphisms of coalgebras for all $p \geq 0$.

EXAMPLE 1.110. Suppose Θ a bialgebra and Ξ a right Θ-Hopf algebra. Given any right $\Theta - \Xi$ module M we can write $TM = M \otimes \Xi$ for the tensor product of the $\Theta - \Xi$ modules M and Ξ, as described in Proposition 1.37. T is a functor from the category of $\Theta - \Xi$ modules to itself. The mappings $d : TM \to M$; $s : TM \to TTM$, defined respectively by $d = M \otimes \epsilon$, $s = M \otimes \psi$ are natural transformations $d : T \to Id$ and $s : T \to TT$. Then $\{T, d, s\}$ is a cotriple on the category of $\Theta - \Xi$ modules. The associated functor W^{Ξ} carries any $\Theta - \Xi$ module M to a simplicial $\Theta - \Xi$ module $W^{\Xi}(M)$ augmenting to M. The contracting homotopies $h : W^{\Xi}_{p-1}(M) \to W^{\Xi}_{p}(M)$ defined by (1.98) are in this context homomorphisms of Θ-modules for all $p \geq 0$. The key point is that the unit $\eta : l \to \Xi$ is a homomorphism of Θ-modules. It follows from Proposition 1.43 that the augmented chain complex $CW^{\Xi}(M) \to M$ is a resolution of the $\Theta - \Xi$ module M by relative projectives, in the sense of Definition 1.45.

12. Simplicial Θ-Coalgebras and Steenrod Operations

In this section we discuss Steenrod operations on cochains, and on cohomology. We define the operations (in a slightly more general context than is classical), and review a formula (1.112) of Kristensen that describes their deviation from additivity at the cochain level. This formula will play an important role when, in Chapter 2, we put Steenrod operations into spectral sequences. We will review the Adem relations, and observe that the Steenrod operations on cohomology, subject to the Adem relations, form an "algebra with coproducts" in the sense of Section 6.

The operations Sq^k were first defined by Steenrod [**83**] on the mod-2 cohomology of a topological space. Later, Dold [**16**] defined them in a more general context, replacing the singular chains on a topological space by an arbitrary simplicial coalgebra over $\mathbb{Z}/2$. The ideas in this section are those of Steenrod and Dold, but we generalize a little further. In place of Dold's simplicial coalgebras over $\mathbb{Z}/2$ we use simplicial coalgebras over Θ, where Θ is an arbitrary (cocommutative) bialgebra over $\mathbb{Z}/2$. For our coefficient ring, in place of $\mathbb{Z}/2$, we use an arbitrary commutative Θ-algebra N. The bialgebra Θ, and all modules, coalgebras, and algebras over Θ, are assumed graded on a fixed abelian group $\mathbb{S}$, a finite product of copies of the integers.

Suppose R a simplicial Θ-module, and N a commutative Θ-algebra. Define a cochain map $\phi : \mathrm{Hom}_{\Theta}(CR, N) \otimes \mathrm{Hom}_{\Theta}(CR, N) \to \mathrm{Hom}_{\Theta}(CR \otimes CR, N)$ to be the diagonal arrow in the following diagram:

(1.99)
$$\begin{array}{ccc} \mathrm{Hom}_{\Theta}(CR, N) \otimes \mathrm{Hom}_{\Theta}(CR, N) & & \\ \downarrow \rho & \searrow \phi & \\ \mathrm{Hom}_{\Theta}(CR \otimes CR, N \otimes N) & \xrightarrow[\mu_*]{} & \mathrm{Hom}_{\Theta}(CR \otimes CR, N). \end{array}$$

Here ρ is as in (1.43a), and μ_* is induced by the product $\mu : N \otimes N \to N$. Thus, given cochains $x, y \in \mathrm{Hom}_{\Theta}(CR, N)$, then the cochain $\phi(x \otimes y)$ is the composition:

(1.100)
$$CR \otimes CR \xrightarrow{x \otimes y} N \otimes N \xrightarrow{\mu} N.$$

We write $T : CR \otimes CR \to CR \otimes CR$ for the transposition mapping, as in Section 9. Since Θ is cocommutative, T is linear over Θ. Since N is commutative,

(1.101)
$$\phi(x \otimes y) = T^* \phi(y \otimes x).$$

Now suppose in addition that R is a simplicial Θ-coalgebra. Then the coproduct $\psi : R \to R \times R$ induces a map of chain complexes over Θ:

$$\psi_* : CR \to C(R \times R), \tag{1.102}$$

and consequently a map of cochain complexes

$$\psi^* : \mathrm{Hom}_\Theta(C(R \times R), N) \to \mathrm{Hom}_\Theta(CR, N). \tag{1.103}$$

By the commutativity of ψ,

$$\psi^* T^* = \psi^*. \tag{1.104}$$

Suppose now we are given a cup-k product $\{D_k \mid k \geq 0\}$ in the sense of Definition 1.91. We can then define a natural map of cochain complexes

$$m : \mathrm{Hom}_\Theta(CR, N) \otimes \mathrm{Hom}_\Theta(CR, N) \to \mathrm{Hom}_\Theta(CR, N) \tag{1.105}$$

by the formula:

$$m = \psi^* D_0^* \phi. \tag{1.106}$$

This cochain operation induces a product

$$\mu : H^{*,*} \mathrm{Hom}_\Theta(CR, N) \otimes H^{*,*} \mathrm{Hom}_\Theta(CR, N) \to H^{*,*} \mathrm{Hom}_\Theta(CR, N) \tag{1.107}$$

by which $H^{*,*} \mathrm{Hom}_\Theta(CR, N)$ becomes a commutative algebra graded on $\mathbb{Z} \times \mathbb{S}$. We define also for all k, n with $0 \leq k \leq n$ and all $s \in \mathbb{S}$ functions

$$S^k : \mathrm{Hom}_\Theta^{n,s}(CR, N) \to \mathrm{Hom}_\Theta^{n+k,2s}(CR, N) \tag{1.108}$$

by the customary formula

$$S^k(x) = \psi^* D_{n-k}^* \phi(x \otimes x) + \psi^* D_{n-k+1}^* \phi(x \otimes \delta x). \tag{1.109}$$

The duals of equations (1.59), together with (1.101) and (1.104) imply that:

$$\delta S^k + S^k \delta = 0. \tag{1.110}$$

S^k is not a homomorphism. Its "deviation from additivity" is a function Δ^k defined for all $x, y \in \mathrm{Hom}_\Theta^n(CR, N)$ by

$$\Delta^k(x, y) = S^k(x + y) - S^k(x) - S^k(y). \tag{1.111}$$

The duals of equations (1.59), together with (1.101) and (1.104) imply that:

$$\Delta^k(x, y) = \psi^*[\delta D_{n-k+2}^* \phi(x \otimes \delta y) + \delta D_{n-k+1}^* \phi(y \otimes x) + D_{n-k+2}^* \phi(\delta x \otimes \delta y)]. \tag{1.112}$$

It follows from (1.110) and (1.112) that for all $k, n \geq 0$ and $s \in \mathbb{S}$, the cochain operation S^k induces an additive cohomology operation:

$$Sq^k : H^{n,s} \mathrm{Hom}_\Theta(CR, N) \to H^{n+k,2s} \mathrm{Hom}_\Theta(CR, N). \tag{1.113}$$

The additivity of the Sq^k is of course in Steenrod [**83**]; (1.112) is from Kristensen [**34**].

PROPOSITION 1.111. *Suppose Θ is a bialgebra over $\mathbb{Z}/2$, and Let R be a simplicial Θ-coalgebra, and N a commutative Θ-algebra. Then the product* (1.107) *and Steenrod squares* (1.113) *are independent of the choice of cup-k product, are natural with respect to maps of the variables R and N, and satisfy the Cartan formula*

$$Sq^k(uv) = \sum_{i+j=k} Sq^i u \cdot Sq^j v \tag{1.114}$$

and the Adem relations

$$Sq^a Sq^b u = \sum_{j\geq 0} \binom{b-1-j}{a-2j} Sq^{a+b-j} Sq^j u; \quad 0 \leq a < 2b \tag{1.115}$$

for all $u, v \in H^{*,*} \operatorname{Hom}_\Theta(CR, N)$.

These results are due to Steenrod, Adem, Cartan, Dold, and May [**2, 3, 13, 16, 48, 83, 85**]. As we remarked at the start of this section, we have replaced their ground ring $\mathbb{Z}/2$ by an arbitrary cocommutative bialgebra Θ over $\mathbb{Z}/2$; and the classical coefficient ring $\mathbb{Z}/2$ by an arbitrary commutative Θ-algebra N. But the original proofs go through without significant change. Notice we make no special statements about Sq^0; in the general context which we consider here Sq^0 is not always the identity operation.

We will write $\mathcal{H}$ for the Steenrod algebra over $\mathbb{Z}/2$, generated as an algebra by the symbols $\{Sq^k \mid k \geq 0\}$ and subject to the Adem relations:

$$Sq^a Sq^b = \sum_{j\geq 0} \binom{b-1-j}{a-2j} Sq^{a+b-j} Sq^j; \quad 0 \leq a < 2b. \tag{1.116}$$

Since we do not regard Sq^0 as the unit, the relations all have length two in the generating set. Consequently we can and will regard $\mathcal{H}$ as an algebra graded on $\mathbb{Z} \times \mathbb{Z}$:

$$\mathcal{H}_{n,v} = \operatorname{Span}\{Sq^{i_1} \dots Sq^{i_v} \mid i_1 \geq 0, \dots, i_v \geq 0; i_1 + \dots + i_v = n\} \tag{1.117}$$

where by definition n is the "total" degree. The following result is proved in [**79**].

PROPOSITION 1.112. *The bigraded Steenrod algebra* $\mathcal{H}$ *has a unique structure as an algebra with coproducts in which:*

1. *An element of* $\mathcal{H}_{n,v}$ *has additive degree* n *and multiplicative degree* v *in the sense of Definition 1.47;*
2. *The coproduct on* $\mathcal{H}_{*,1}$ *is given by:*

$$\psi(Sq^k) = \sum_{i+j=k} Sq^i \otimes Sq^j \tag{1.118}$$

for each $k \geq 0$.

Now from Proposition 1.111 we have:

PROPOSITION 1.113. *Let* Θ *be an* $\mathbb{S}$*-graded bialgebra over the ground field* $\mathbb{Z}/2$. *Suppose* R *a simplicial object over the category of* Θ*-coalgebras, and* N *a commutative* Θ*-algebra. Then with the action of* Sq^k *as in* (1.113), *the* $(\mathbb{Z} \times \mathbb{S})$*-graded cohomology group* $H^{*,*} \operatorname{Hom}_\Theta(CR, N)$ *becomes a left module over the algebra with coproducts* $\mathcal{H}$, *in the sense of Definition 1.50. Here an element of* $H^{n,s} \operatorname{Hom}_\Theta(CR, N)$ *has additive degree* n *and multiplicative degree* s *in the language of Definition 1.50. Further, with the product as in* (1.107), $H^{*,*} \operatorname{Hom}_\Theta(CR, N)$ *becomes a commutative left algebra over the algebra with coproducts* $\mathcal{H}$ *in the sense of Remark 1.54.*

We close this section with a remark about Sq^0. Although Sq^0 is not in general the identity operator, it can be described at the cochain level in a particularly simple way.

PROPOSITION 1.114. *Suppose $u \in H^{n,*}\operatorname{Hom}_\Theta(CR, N)$ is represented by the Θ-linear cocycle $x : C_nR \to N$. Then the cocycle $\overline{x} : C_nR \to N$ defined for all $\sigma \in R_n$ by:*

$$\overline{x}(\sigma) = x(V\sigma) \cdot x(V\sigma) \tag{1.119}$$

represents Sq^0u. Here V is the Verschiebung on the coalgebra R_n.

PROOF. By definition, Sq^0u is represented by the cocycle $\overline{x} : C_nR \to N$ which is the composition:

$$C_nR \xrightarrow{\psi_*} C_n(R \times R) \xrightarrow{D_n} (CR \otimes CR)_{2n} \xrightarrow{x \otimes x} N \otimes N \xrightarrow{\mu} N. \tag{1.120}$$

Here we may assume that we have chosen a cup-k product $\{D_k \mid k \geq 0\}$ that is special, in the sense of Definition 1.94. Now given $\sigma \in R_n$ we have as in (1.17) an expansion

$$\psi(\sigma) = \sum_i (\sigma_i \otimes \sigma_i' + \sigma_i' \otimes \sigma_i) + V\sigma \otimes V\sigma.$$

So by the second assertion of Proposition 1.96 we have:

$$D_n\psi(\sigma) = \sum_i (\sigma_i \otimes \sigma_i' + \sigma_i' \otimes \sigma_i) + V\sigma \otimes V\sigma.$$

(1.119) now follows from the commutativity of the algebra N. □

13. Steenrod Operations on the Cohomology of Simplicial Sets

In this section and the next we will give examples of contexts in which the theory of the previous section applies, and actions of the Steenrod algebra are defined.

Given a simplicial set $\mathcal{R}$ we write R for the simplicial vector space over the ground field l of which $\mathcal{R}$ is the basis. Then R becomes a simplicial l-coalgebra as described in Section 10.

Now we take $l = \mathbb{Z}/2$ and apply the constructions of Section 12 to the simplicial $\mathbb{Z}/2$-coalgebra R. Here we are taking the bialgebra Θ of Section 12 to be just $\mathbb{Z}/2$, and the grading group $\mathbb{S}$ to be 0. We take the coefficient ring to be $N = \mathbb{Z}/2$. Then (1.107) and (1.113) give, respectively, product and Steenrod operations on the ordinary mod-2 cohomology of $\mathcal{R}$:

$$H^*\mathcal{R} \otimes H^*\mathcal{R} \to H^*\mathcal{R} \tag{1.121}$$

$$Sq^k : H^n\mathcal{R} \to H^{n+k}\mathcal{R}. \tag{1.122}$$

The product is the cup-product as originally defined by Whitney on the cohomology of a topological space. The product (1.78) is the negative of this one.

The squaring operations are those originally defined by Steenrod. From the earliest work by Steenrod it is known that Sq^0 of (1.122) is the identity operation. In fact this can be seen as a consequence of Proposition 1.114, together with the facts that the Verschiebung on R is the identity mapping, and $\alpha^2 = \alpha$ for all α in the coefficient ring $\mathbb{Z}/2$. So in this case the Cartan formula (1.114) and Adem relations (1.115) are to be interpreted with $Sq^0 = 1$. The resulting algebra of operations on the cohomology of simplicial sets is again generated by the Sq^k and subject to the formal relations (1.116), with $Sq^0 = 1$. This is the classical Steenrod algebra $\boldsymbol{\mathcal{A}}$, known from the work of Milnor [**52**] to be a Hopf algebra, with coproduct determined by (1.118). Equations (1.114),(1.115) now say that if $\mathcal{R}$ is a simplicial set, then its $\mathbb{Z}/2$-cohomology ring is a commutative left $\boldsymbol{\mathcal{A}}$-algebra.

REMARK 1.115. Suppose $\mathcal{R}$ and $\mathcal{S}$ are simplicial sets. From the fact that $H^*(\mathcal{R} \times \mathcal{S})$ is an $\boldsymbol{\mathcal{A}}$-algebra, it follows easily that the cross product (1.71) is a homomorphism of $\boldsymbol{\mathcal{A}}$-modules.

REMARK 1.116. In our work on the Eilenberg-Moore spectral sequence in Chapter 7 we will need to consider the homology of a simplicial set $\mathcal{R}$ as a right module over $\overline{\boldsymbol{\mathcal{A}}}$, the negative of the Steenrod algebra. This is possible if the mod-2 homology of $\mathcal{R}$ is finite in each degree ("of finite type"). In that case the map $j_{\mathcal{R}} : H_*\mathcal{R} \to \operatorname{Hom}(H^*\mathcal{R}, \mathbb{Z}/2)$ of (1.66) is an isomorphism of $\mathbb{Z}/2$-vector spaces. But by the (negative of) construction (1.25), $\operatorname{Hom}(H^*\mathcal{R}, \mathbb{Z}/2)$ becomes a right $\overline{\boldsymbol{\mathcal{A}}}$-module. So we use the isomorphism $j_{\mathcal{R}}$ to define $H_*\mathcal{R}$ as a right $\overline{\boldsymbol{\mathcal{A}}}$-module.

PROPOSITION 1.117. *Let $\mathcal{R}$ and $\mathcal{S}$ be simplicial sets, both with mod-2 homology finite in each degree. Then the pairing $H_*(\mathcal{R} \times \mathcal{S}) \xrightarrow{H_*f} H_*\mathcal{R} \otimes H_*\mathcal{S}$ of* (1.69) *is an isomorphism of right $\overline{\boldsymbol{\mathcal{A}}}$-modules.*

PROOF. Under current hypotheses all the mappings in (1.72) are isomorphisms of vector spaces; so if we can show that $j_{\mathcal{R}\times\mathcal{S}}$, $j_{\mathcal{R}} \otimes j_{\mathcal{S}}$, cr^*, and ν are linear over $\overline{\boldsymbol{\mathcal{A}}}$ then we are done. But the first two of these are $\overline{\boldsymbol{\mathcal{A}}}$-linear by definition, and cr^* is $\overline{\boldsymbol{\mathcal{A}}}$-linear by Remark 1.115. Finally ν is linear over $\overline{\boldsymbol{\mathcal{A}}} \otimes \overline{\boldsymbol{\mathcal{A}}}$ by the negative of Proposition 1.22; so it is also linear over $\overline{\boldsymbol{\mathcal{A}}}$. □

COROLLARY 1.118. *Let $\mathcal{R}$ be a simplicial set with mod-2 homology finite in each degree. Then $H_*\mathcal{R}$ is an $\overline{\boldsymbol{\mathcal{A}}}$-coalgebra.*

PROOF. This follows from Proposition 1.117 and the description (1.77) of the coproduct on $H_*\mathcal{R}$. □

Similarly we have:

COROLLARY 1.119. *Let $\mathcal{G}$ be a simplicial group and $\mathcal{E}$ a right $\mathcal{G}$-space. Suppose the mod-2 homologies of $\mathcal{G}$ and $\mathcal{E}$ of finite type. Let $H_*\mathcal{G}$ be considered a Hopf algebra, and $H_*\mathcal{E}$ a right $H_*\mathcal{G}$-module, as described in Section 10. Then $H_*\mathcal{G}$ is a right $\overline{\boldsymbol{\mathcal{A}}}$-Hopf algebra, and $H_*\mathcal{E}$ is a right $\overline{\boldsymbol{\mathcal{A}}} - H_*\mathcal{G}$ module.*

PROOF. These results follow from Proposition 1.117, and from the descriptions (1.79), (1.82) of the product on $H_*\mathcal{G}$, and of the action of $H_*\mathcal{G}$ on $H_*\mathcal{E}$. □

14. Steenrod Operations on the Cohomology of Hopf Algebras

In this section we give a second application of the general theory of Steenrod operations set out in Section 12: we review the definition of the operations on the cohomology of a cocommutative Hopf algebras Θ with coefficients in a Θ-coalgebra M. These operations are due to Liulevicius ([**40**]). We give two equivalent definitions. One is expressed in terms of a specific resolution of M as a Θ-module: a resolution made from the chains on the particular simplicial coalgebra $W^{\Theta}(M)$. The second definition is more flexible, and is expressed in terms of an arbitrary projective resolution of M. We will have uses for both.

Consider an arbitrary cocommutative Hopf algebra Θ over $\mathbb{Z}/2$, graded on an abelian group $\mathbb{S}$, and suppose given a right Θ-coalgebra M. Then we have $W^{\Theta}(M)$, the simplicial object over the category of Θ-coalgebras described in Example 1.109. If N is any commutative Θ-algebra, the constructions (1.107), (1.113) give a product

$$H^{*,*}\operatorname{Hom}_\Theta(CW^\Theta(M),N)\otimes H^{*,*}\operatorname{Hom}_\Theta(CW^\Theta(M),N)\to H^{*,*}\operatorname{Hom}_\Theta(CW^\Theta(M),N) \tag{1.123}$$

and Steenrod operations:

$$Sq^k : H^{n,s}\operatorname{Hom}_\Theta(CW^\Theta(M),N)\to H^{n+k,2s}\operatorname{Hom}_\Theta(CW^\Theta(M),N). \tag{1.124}$$

But we observed in Section 11 that $CW^\Theta(M)$ can be viewed as a free resolution of M as a Θ-module, so that $H^{*,*}\operatorname{Hom}_\Theta(CW^\Theta(M),N)=\operatorname{Ext}^{*,*}_\Theta(M,N)$. So we have from Proposition 1.113:

PROPOSITION 1.120. *Let Θ be a Hopf algebra over $\mathbb{Z}/2$, M a right Θ-coalgebra and N a commutative right Θ-algebra. Then equations* (1.123) *and* (1.124) *define a product*

$$\operatorname{Ext}^{*,*}_\Theta(M,N)\otimes\operatorname{Ext}^{*,*}_\Theta(M,N)\to\operatorname{Ext}^{*,*}_\Theta(M,N) \tag{1.125}$$

and Steenrod operations

$$Sq^k : \operatorname{Ext}^{n,s}_\Theta(M,N)\to\operatorname{Ext}^{n+k,2s}_\Theta(M,N) \tag{1.126}$$

under which $\operatorname{Ext}^{*,*}_\Theta(M,N)$ *becomes a commutative left algebra over the algebra with coproducts* $\mathcal{H}$.

(1.125) is the well-known cup product in the cohomology of a cocommutative Hopf algebra. (1.126) are the Steenrod squares in Ext originally constructed by Liulevicius [**40**] in the case $M=N=\mathbb{Z}/2$.

In some circumstances it is convenient to have a more flexible definition of the operations (1.125) and (1.126), one which replaces the particular resolution $CW^\Theta(M)$ by an arbitrary projective resolution of M. Indeed, this was the approach originally taken in [**40**]. Given a right Θ-coalgebra M, let $\cdots\to\mathcal{P}_n(M)\to\cdots\to\mathcal{P}_0(M)\to M\to 0$ be a projective resolution of M as a Θ-module, with degree-preserving differentials. We will write $\mathcal{P}(M)$ for the complex: $\cdots\to\mathcal{P}_n(M)\to\cdots\to\mathcal{P}_0(M)$, and $\mathcal{P}(M)\to M\to 0$ for the augmented version.

DEFINITION 1.121. By a homological cup-k product for the projective resolution $\mathcal{P}(M)$ we mean a family of Θ-linear maps

$$\mathcal{D}_k:\mathcal{P}_n(M)\to(\mathcal{P}(M)\otimes\mathcal{P}(M))_{n+k}$$

defined for all $n,k\geq 0$, preserving internal degree, and having the properties:

1. $\mathcal{D}_0$ is a chain map carrying the coproduct on M;
2. For each $k\geq 1$:

$$\partial\mathcal{D}_k+\mathcal{D}_k\partial=\mathcal{D}_{k-1}+T\mathcal{D}_{k-1}. \tag{1.127}$$

Indeed, the standard arguments of homological algebra imply the existence of a Θ-linear chain map $\mathcal{D}_0:\mathcal{P}(M)\to\mathcal{P}(M)\otimes\mathcal{P}(M)$ that carries the Θ-linear coproduct $\psi:M\to M\otimes M$. This $\mathcal{D}_0$ is unique up to Θ-linear chain homotopy. But since the coproduct on Θ is commutative, the switching map $T:\mathcal{P}(M)\otimes\mathcal{P}(M)\to\mathcal{P}(M)\otimes\mathcal{P}(M)$ is a homomorphism of Θ-modules; and since the coproduct on M is commutative, the mapping $T\mathcal{D}_0$, like $\mathcal{D}_0$, is a Θ-linear chain map that carries $\psi:M\to M\otimes M$. Consequently there must be a Θ-linear chain homotopy between $\mathcal{D}_0$ and $T\mathcal{D}_0$. This is $\mathcal{D}_1$. One now proceeds by induction on k, using analogous arguments, to construct the full sequence of homomorphisms $\{\mathcal{D}_k \mid k \geq 0\}$ having the properties listed in Definition 1.121. The argument is now standard, and can be found in [**40**].

If N is any right Θ-module, each $\mathcal{D}_k$ induces an $\mathbb{Z}/2$-linear mapping which lowers homological degree by k:

$$\mathcal{D}_k^* : \mathrm{Hom}_\Theta^{n+k}(\mathcal{P}(M) \otimes \mathcal{P}(M), N) \to \mathrm{Hom}_\Theta^n(\mathcal{P}(M), N). \tag{1.128}$$

defined for all $n \geq 0$. These maps satisfy the duals of (1.127). If in addition N is a commutative Θ-algebra, we have by analogy with (1.100) a mapping of cochain complexes

$$\phi : \mathrm{Hom}_\Theta(\mathcal{P}(M), N) \otimes \mathrm{Hom}_\Theta(\mathcal{P}(M), N) \to \mathrm{Hom}_\Theta(\mathcal{P}(M) \otimes \mathcal{P}(M), N) \tag{1.129}$$

defined by setting $\phi(x \otimes y)$ equal to the compostion:

$$\mathcal{P}(M) \otimes \mathcal{P}(M) \xrightarrow{x \otimes y} N \otimes N \xrightarrow{\mu} N. \tag{1.130}$$

This mapping satisfies (1.101). We now define a map of cochain complexes

$$m : \mathrm{Hom}_\Theta(\mathcal{P}(M), N) \otimes \mathrm{Hom}_\Theta(\mathcal{P}(M), N) \to \mathrm{Hom}_\Theta(\mathcal{P}(M), N)$$

by the customary formula:

$$m = \mathcal{D}_0^* \phi. \tag{1.131}$$

We define also for all k, n with $0 \leq k \leq n$ and all $s \in \mathbb{S}$ the functions

$$S^k : \mathrm{Hom}_\Theta^{n,s}(\mathcal{P}(M), N) \to \mathrm{Hom}_\Theta^{n+k,2s}(\mathcal{P}(M), N)$$

by setting

$$S^k(x) = \mathcal{D}_{n-k}^* \phi(x \otimes x) + \mathcal{D}_{n-k+1}^* \phi(x \otimes \delta x) \tag{1.132}$$

for each $x \in \mathrm{Hom}_\Theta^n(\mathcal{P}(M), N)$. Then by (1.101) and the dual of (1.127):

$$\delta S^k + S^k \delta = 0.$$

Then the cochain operations m and S^k define respectively a product and Steenrod squares as in (1.125) and (1.126). Liulevicius shows in [**40**] that these operations are independent of the choice of projective resolution $\mathcal{P}(M)$ of M, and independent of the choice of the homological cup-k product $\{\mathcal{D}_k \mid k \geq 0\}$.

It is easy to see that the definitions of product and Steenrod squares that we have just given, in the language of projective resolutions, agree with the definitions we originally gave in terms of simplicial objects. One need only observe that if $\{D_k \mid k \geq 0\}$ is a cup-k product in the sense of Definition 1.91, and if M is a right Θ-coalgebra, we can define for all pairs of integers $n, k \geq 0$ a Θ-linear homomorphism $\mathcal{D}_k : C_n W^\Theta(M) \to (CW^\Theta(M) \otimes CW^\Theta(M))_{n+k}$ by setting:

$$\mathcal{D}_k = D_k \psi_*.$$

Here $\psi_* : CW^\Theta(M) \to C(W^\Theta(M) \times W^\Theta(M))$ is the mapping of chain complexes over Θ that is induced by the coproduct on $W^\Theta(M)$, as in (1.102). But as we observed in Section 11, $CW^\Theta(M)$ is a projective resolution of M as a Θ-module; and it is clear that the family $\{\mathcal{D}_k \mid k \geq 0\}$ satisfies the conditions 1-4 of Defintion 1.121. So this family forms a "homological cup-k product" for the resolution $CW^\Theta(M)$. Thus, the cochain operations (1.106) and (1.131) agree; and the cochain operations (1.109) and (1.132) agree. So the corresponding cohomology operations agree as well.

REMARK 1.122. Suppose Θ acts trivially on the algebra N, and suppose that both Θ and M are locally finite in the sense of Definition 1.1. Suppose further that both Θ and M are bounded below, and that N is bounded above, in the sense of Definition 1.8. Then one can choose a projective resolution $\mathcal{P}(M)$ of M as a right Θ-module, in such a way that each Θ-module $\mathcal{P}_n(M)$ is locally finite and bounded below. Then it follows from Remark 1.20 that one has an isomorphism of $\mathbb{S}$-graded vector spaces

$$\mathrm{Ext}_\Theta^{n,*}(M, N) = \mathrm{Ext}_\Theta^{n,*}(M, \mathbb{Z}/2) \otimes \overline{N}^* \tag{1.133}$$

for each $n \geq 0$. We can also regard $\overline{N}$ as graded on $\mathbb{Z} \times \mathbb{S}$, with $\overline{N}^{0,s} = \overline{N}^s$ and $\overline{N}^{n,s} = 0$ if $n > 0$. Then it follows easily from our definitions that we have as well an isomorphism of $(\mathbb{Z}\times\mathbb{S})$-graded algebras: $\mathrm{Ext}_\Theta^{*,*}(M, N) = \mathrm{Ext}_\Theta^{*,*}(M, \mathbb{Z}/2) \otimes \overline{N}^{*,*}$; and further that under this identification the Steenrod squares (1.126) are given by

$$Sq^k(u \otimes v) = Sq^k u \otimes v^2$$

for all $u \in \mathrm{Ext}_\Theta^{*,*}(M, \mathbb{Z}/2)$, $v \in \overline{N}^{*,*}$.

15. Bisimplicial Objects

This section will give our conventions concerning bisimplicial objects, their augmentations, and associated diagonal objects.

We will be working with bisimplicial objects $X = \{X_{p,q} \,|\, p \geq 0, q \geq 0\}$ over a category $\mathcal{C}$, throughout this book. We write d_i^h, s_i^h for the "horizontal" face and degeneracy operators: $d_i^h : X_{p,q} \to X_{p-1,q}$; $s_i^h : X_{p,q} \to X_{p+1,q}$; and d_i^v, s_i^v for the "vertical" operators. To any bisimplicial object over a category $\mathcal{C}$ is associated a simplicial object over $\mathcal{C}$ called $\mathrm{Diag}\, X$. It is defined by $(\mathrm{Diag}\, X)_n = X_{n,n}$; the face and degeneracy operators are $d_i = d_i^h d_i^v$; $s_i = s_i^h s_i^v$.

DEFINITION 1.123. If X is a bisimplicial object over a category $\mathcal{C}$, by an augmentation of X we will mean a pair $\{\lambda, R\}$, where R is a simplicial object over $\mathcal{C}$, and $\lambda : X_{0,*} \to R_*$ is a morphism of simplicial $\mathcal{C}$-objects satisfying:

$$\lambda d_1^h = \lambda d_0^h. \tag{1.134}$$

Fix a bialgebra Θ over a field l, and graded on an abelian group $\mathbb{S}$. We will be working with bisimplicial objects over the category of Θ-modules ("bisimplicial Θ-modules") and with bisimplicial objects over the category of Θ-coalgebras ("bisimplicial Θ-coalgebras"). All face and degeneracy operators preserve internal degree. In the ensuing discussion we have ruthlessly suppressed internal degree from the notation; writing explicitly only those gradings arising from the bisimplicial structure. If X is a bisimplicial Θ-module we write CX for the associated double chain complex of Θ-modules defined by $C_{p,q}X = X_{p,q}$. The horizontal differential $\partial^h : C_{p,q}X \to C_{p-1,q}X$ is $\partial^h = \sum_{0\leq i\leq p}(-1)^i d_i^h$; the vertical differential $\partial^v : C_{p,q}X \to C_{p,q-1}X$ is $\partial^v = \sum_{0\leq i\leq q}(-1)^i d_i^v$. We will also write CX for the "total" chain complex associated with X. This is the ordinary chain complex of Θ-modules with $C_nX = \bigoplus_{p+q=n} X_{p,q}$ and total differential $\partial u = \partial^h u + (-1)^p \partial^v u$ for all $u \in X_{p,q}$. Whenever we write a double chain complex with a single subscript, the subscript refers to total degree.

If a bisimplicial Θ-module X is equipped with an augmentation $\{\lambda, R\}$, then it follows from (1.134) that λ defines a chain mapping $C\lambda : CX \to CR$ from the total chains on X to the chains on R. For $x \in X_{p,q}$ this is defined by:

$$C\lambda(x) = \begin{cases} \lambda(x) & \text{if } p = 0 \\ 0 & \text{if } p > 0. \end{cases} \tag{1.135}$$

It is classical that if X is a bisimplicial Θ-module, the chain complex on the diagonal $C(\operatorname{Diag} X)$ is naturally chain-homotopy equivalent to the total chain complex CX. In Chapter 6 we will need a specific formula for the chain equivalence, which we now review. Write F^v for the vertical "front face" operator on a bisimplicial Θ-module X, and $\tilde{F}^h$ for the horizontal "back face" operator. For $a \in X_{p,q}$ these are defined by: $F^v(a) = d_0^v(a)$ and $\tilde{F}^h(a) = d_p^h(a)$. Define a Θ-linear function $f : C(\operatorname{Diag} X) \to CX$ by writing for $a \in X_{n,n} = C_n(\operatorname{Diag} X)$:

$$f(a) = \sum_{i=0}^{n} (\tilde{F}^h)^{n-i}(F^v)^i(a). \tag{1.136}$$

When X is determined by a pair of simplicial Θ-modules R, S according to the formula $X_{p,q} = R_p \otimes S_q$, then (1.136) reduces to (1.53).

PROPOSITION 1.124. *Let Θ be a Hopf algebra, and X a bisimplicial Θ-module. Then the mapping $f : C(\operatorname{Diag} X) \to CX$ defined by (1.136) is a natural chain-homotopy equivalence of chain complexes over Θ.*

This is proved as in [**19, 20**]. Eilenberg and Mac Lane deal with the case in which $\Theta = l$, and the bisimplicial object X is determined by a pair of simplicial l-modules R, S as above. But their arguments extend easily to the more general case we consider here.

Supposing X a bisimplicial Θ-module and N a Θ-module, we define a double cochain complex over the category of $\mathbb{S}$-graded vector spaces, $\operatorname{Hom}_\Theta^{*,*}(CX, N)$, by writing $\operatorname{Hom}_\Theta^{p,q}(CX, N) = \operatorname{Hom}_\Theta(X_{p,q}, N)$. The "horizontal" differential

$$\delta_h : \operatorname{Hom}_\Theta^{p-1,q}(CX, N) \to \operatorname{Hom}_\Theta^{p,q}(CX, N)$$

is the dual of the mapping ∂^h defined above; and the "vertical" differential

$$\delta_v : \operatorname{Hom}_\Theta^{p,q-1}(CX, N) \to \operatorname{Hom}_\Theta^{p,q}(CX, N)$$

is dual to ∂^v.

16. The Spectral Sequence of a Bisimplicial Θ-Module

In this section we review the definition of the first-quadrant cohomology spectral sequence associated with a bisimplicial Θ-module. We give the definition in a form that we can apply easily in Chapter 2, when we define Steenrod operations in the spectral sequence.

Given a bialgebra Θ and a bisimplicial Θ-module X, filter the n-dimensional chains of the total complex CX by writing:

$$F_pC_nX = \bigoplus_{i=0}^{p} X_{i,n-i} \tag{1.137}$$

for all $p \geq 0$. Set $F_pC_nX = 0$ if $p < 0$. For each integer p the differential ∂ on the total chains satisfies $\partial(F_pC_nX) \subseteq F_pC_{n-1}X$ for all $n > 0$, so we get a sub-chain complex F_pCX of the total chains CX by writing $(F_pCX)_n = F_pC_nX$ for each $n \geq 0$. So we have given the chain complex CX an ascending filtration: $\cdots F_{p-1}CX \subseteq F_pCX \cdots$.

If now we are given a Θ-module N we write $\operatorname{Hom}^n_\Theta(CX, N) = \operatorname{Hom}_\Theta(C_nX, N)$, as in (1.39). These vector spaces make up a cochain complex $\operatorname{Hom}_\Theta(CX, N)$. Corresponding to (1.137) there is a dual, descending filtration on this cochain complex given by:

$$F^p \operatorname{Hom}_\Theta(CX, N) = [\, x \in \operatorname{Hom}_\Theta(CX, N) \mid x(F_{p-1}CX) = 0\,]. \tag{1.138}$$

Consequently the cohomology of this cochain complex also has a descending filtration, given by:

(1.139)
$$F^p H^* \operatorname{Hom}_\Theta(CX, N) = \operatorname{image}[H^* F^p \operatorname{Hom}_\Theta(CX, N) \to H^* \operatorname{Hom}_\Theta(CX, N)].$$

The associated bigraded vector space $E_0 H^* = \{E_0^{p,q} H^* \mid p \geq 0,\, q \geq 0\}$, is by definition:

$$E_0^{p,q} H^* = \frac{F^p H^{p+q} \operatorname{Hom}_\Theta(CX, N)}{F^{p+1} H^{p+q} \operatorname{Hom}_\Theta(CX, N)}. \tag{1.140}$$

REMARK 1.125. In the work that follows we will be dealing with cochain complexes C^* and cohomology groups H^* that are equipped with descending filtrations: $\cdots F^pC^* \subseteq F^{p-1}C^* \subseteq \cdots \subseteq F^0C^* = C^*$; $\cdots F^pH^* \subseteq F^{p-1}H^* \subseteq \cdots \subseteq F^0H^* = H^*$. In such cases we will always take $F^pC^* = C^*$ if $p < 0$ and $F^pH^* = H^*$ if $p < 0$. We will often write F^p in place of F^pC^* or F^pH^* when no confusion can arise.

If Θ is a bialgebra, N a Θ-module, and X a bisimplicial Θ-module, then we will refer to "the spectral sequence of the bisimplicial Θ-module X with coefficients in N". This is just the spectral sequence associated with the filtered cochain complex (1.138). We define this in the usual way. For all integers p, q, r set:

$$Z_r^{p,q} = [x \in F^p \operatorname{Hom}^{p+q}_\Theta(CX, N) \mid \delta x \in F^{p+r} \operatorname{Hom}_\Theta(CX, N)]. \tag{1.141}$$

Also write $Z_r^{p,*}$ for the graded vector space: $Z_r^{p,*} = \{\, Z_r^{p,q} \mid q \geq 0\}$. Notice that $Z_r^{p,q} = 0$ if $q < 0$, but $Z_r^{p,q}$ can be interesting if $p < 0$ (recall Remark 1.125). Clearly, for all p, q, r we have:

$$Z_r^{p+1,*} \subseteq Z_{r+1}^{p,*} \quad \text{and} \quad Z_{r+1}^{p,*} \subseteq Z_r^{p,*}. \tag{1.142}$$

Now we define the sub-vector space $B_r^{p,q} \subseteq Z_r^{p,q}$ by setting:

$$B_r^{p,q} = \delta Z_{r-1}^{p-r+1,q+r-2} + Z_{r-1}^{p+1,q-1}. \tag{1.143}$$

The bigraded E_r term of the spectral sequence is given as usual by:

$$E_r^{p,q} = Z_r^{p,q} / B_r^{p,q}. \tag{1.144}$$

The definitions imply that $E_r^{p,q} = 0$ if either $p < 0$ or $q < 0$. The total differential δ of the cochain complex $\operatorname{Hom}_\Theta(CX, N)$ induces the differentials of the spectral sequence:

$$d_r : E_r^{p,q} \to E_r^{p+r,q-r+1}. \tag{1.145}$$

As usual, we have that E_{r+1} is the homology of E_r under this differential:

$$E_{r+1} = H(E_r; d_r), \tag{1.146}$$

and that

$$E_r^{p,q} = E_{r+1}^{p,q} = \cdots = E_\infty^{p,q} \quad \text{if} \quad r > \max(p, q+1), \tag{1.147}$$

where one defines $E_\infty^{p,q}$ to be the common values of $E_r^{p,q}$ for r sufficiently large. We define an isomorphism of bigraded vector spaces

$$\pi : E_\infty \simeq E_0 H^* \tag{1.148}$$

as follows. Given $\alpha \in E_r^{p,q}$ for $r > \max(p, q+1)$, choose a cochain $x \in Z_r^{p,q}$ that represents α. Since $r > q+1$, x is a cocycle in $F^p \operatorname{Hom}_\Theta^{p+q}(CX, N)$, and represents a cohomology class $[x] \in F^p H^{p+q} \operatorname{Hom}_\Theta(CX, N)$. Abusing notation a little we also write $[x]$ for the corresponding class in $E_0^{p,q} H^*$, and set $\pi(\alpha) = [x]$. It is easy to show that π is well-defined, and an isomorphism. In this sense, the spectral sequence of the bisimplicial Θ-module X with coefficients in N "converges to $H^* \operatorname{Hom}_\Theta(CX, N)$".

Finally we recall that, while (1.144) gives one useful description of E_2, another is given by

$$E_2^{p,q} = H_h^p H_v^q, \tag{1.149}$$

"the horizontal cohomology of the vertical cohomology". Thus, with reference to the horizontal and vertical differentials δ_h, δ_v defined at the end of the last section, the right-hand side of (1.149) is the p'th homology group of the complex:

$$\cdots \to H^q(\operatorname{Hom}_\Theta(C_{p,*}X, N)) \xrightarrow{\delta_h} H^q(\operatorname{Hom}_\Theta(C_{p+1,*}X, N)) \to \cdots.$$

So an element of $E_2^{p,q}$ is represented by a Θ-linear cochain $x : X_{p,q} \to N$ satisfying $\delta_v x = 0$, and for which there exists a Θ-linear cochain $x' : X_{p+1,q-1} \to N$ satisfying $\delta_h x = \delta_v x'$.

17. Cup-k Products and Bisimplicial Θ-Modules

Cup-k products can be used in various ways to define mappings among the chain complexes associated with bisimplicial Θ-modules. In this section we give some of these definitions. We will use them in Chapter 2 and in Chapter 7. This section ends with Proposition 1.126, which will be used only in Chapter 7, when we discuss the Eilenberg-Moore spectral sequence.

Fix a bialgebra Θ. Suppose $\{D_k \mid k \geq 0\}$ is a cup-k product, and suppose R, S simplicial Θ-modules. For any pair of integers (l, m) for which $l + m = p + k$ we write $D_k^{l,m}(R, S) : R_p \otimes S_p \to R_l \otimes S_m$ for the Θ-linear mapping:

$$R_p \otimes S_p \xrightarrow{D_k(R,S)} (CR \otimes CS)_{p+k} \xrightarrow{\text{proj}} R_l \otimes S_m. \tag{1.150}$$

Similarly, suppose X, Y a pair of bisimplicial Θ-modules. For fixed non-negative integers l, m, p, k for which $l + m = p + k$ we define a map of Θ-modules:

$$J_k^{l,m}(X, Y) : C(X_{p,*}) \otimes C(Y_{p,*}) \to C(X_{l,*}) \otimes C(Y_{m,*}).$$

The restriction of $J_k^{l,m}$ to an arbitrary summand $X_{p,a} \otimes Y_{p,c}$ is $D_k^{l,m}(X_{*,a}, Y_{*,c})$; i.e., the composition:

$$
\begin{array}{ccc}
X_{p,a} \otimes Y_{p,c} \subseteq C_p(X_{*,a} \times Y_{*,c}) & \xrightarrow{D_k(X_{*,a}, Y_{*,c})} & (C(X_{*,a}) \otimes C(Y_{*,c}))_{p+k} \\
& & \Big\downarrow \text{proj} \\
& & X_{l,a} \otimes Y_{m,c}.
\end{array}
\tag{1.151}
$$

For each fixed quadruple l, m, p, k, we claim that $J_k^{l,m}(X, Y)$ is a chain map, from the tensor product of chain complexes $C(X_{p,*}) \otimes C(Y_{p,*})$ to the tensor product of chain complexes $C(X_{l,*}) \otimes C(Y_{m,*})$. In fact this follows easily from the naturality of the mapping D_k: in (1.151) the cup-k product is taken with respect to the horizontal simplicial operators in X and Y, so it commutes with the vertical face operators on both $X_{p,*}$ and $Y_{p,*}$. Then we ask, what map in homology is induced by $J_k^{l,m}(X, Y)$? To answer, we define first for each bisimplicial Θ-module X and each $a \geq 0$ a simplicial object $H_a^v(X)$ over the category of Θ-modules. This "vertical homology of X" is defined by

$$H_a^v(X)_p = H_a C(X_{p,*})$$

for each $p \geq 0$. The simplicial operators on $H_a^v(X)$ are then defined in the obvious way by the horizontal simplicial operators on X.

PROPOSITION 1.126. *Suppose given a pair of bisimplicial Θ-modules X, Y and non-negative integers l, m, p, k for which $l + m = p + k$. Then the homomorphism $J_k^{l,m}(X, Y) : C(X_{p,*}) \otimes C(Y_{p,*}) \to C(X_{l,*}) \otimes C(Y_{m,*})$ is a chain map, and the induced map in homology, when restricted to $H_a C(X_{p,*}) \otimes H_c C(Y_{p,*})$, is given by the homomorphism of Θ-modules*

$$D_k^{l,m}(H_a^v X, H_c^v Y) : H_a C(X_{p,*}) \otimes H_c C(Y_{p,*}) \to H_a C(X_{l,*}) \otimes H_c C(Y_{m,*}) \tag{1.152}$$

for all $a, c \geq 0$.

This follows easily from the naturality of the cup-k product.

CHAPTER 2

The Spectral Sequence of a Bisimplicial Coalgebra

Our goal in this chapter is to define Steenrod squaring operations on a class of spectral sequences sufficiently general to include all the special cases we are interested in. The spectral sequences we will work with are those arising from the standard filtration of the cochains on a bisimplicial coalgebra. The Serre spectral sequence for the cohomology of a fiber space, the Eilenberg-Moore spectral sequence for the cohomology of a classifying space (or more generally, of a Borel construction), and the change-of-rings spectral sequence for the cohomology of an extension of Hopf algebras can all be obtained within this context, so our theory of Steenrod operations will apply in each case.

The following are the main results of this chapter. Theorem 2.15 defines Steenrod operations on each page of the spectral sequence. Theorem 2.16 shows that the operations on each page are determined by the operations at the E_2 level. Theorem 2.17 describes how the operations commute with the differentials of the spectral sequence. Theorem 2.22 shows how the operations on E_∞ are compatible with the operations on the cohomology of the total complex. Theorem 2.23 describes the operations on E_2.

We wish to draw the reader's attention to Theorem 2.21, which says simply that if α is an element of the E_r-term of the spectral sequence, then α^2 survives to E_{2r-1}. This result is used in an essential way by Palmieri in [**64**]. The theorem seems especially interesting because, although it is stated purely in terms of the product structure of the spectral sequence, it does not seem to be provable within that context. For the proof one must know something about the "cup-1" product, and how it treats filtration.

In Section 4 of this chapter we give a description of the diagonal squares at the E_2-level, in the case in which the bisimplicial coalgebra is equipped with an action of a simplicial group. This is the most technical section of the chapter. It is needed only for the calculation in Chapter 7 of the diagonal squares on the E_2-term of the Eilenberg-Moore spectral sequence.

In the final two sections of this chapter we sketch applications. In Section 5 we cite work of Andreas Dress [**17**], in which he obtains the Serre spectral sequence as the spectral sequence of a bisimplicial coalgebra. Our theory applies, and gives Steenrod operations in the Serre spectral sequence. In this way we recover the results of Vasquez-Garcia, Araki, and Kristensen, [**6, 34, 87**]. In Section 6 we describe work of Paul Goerss [**25**], who uses Steenrod operations in a spectral sequence to compute the algebra of operations on the André-Quillen cohomology of simplicial augmented algebras.

1. Bisimplicial Θ-Coalgebras

Fix a cocommutative bialgebra Θ over $\mathbb{Z}/2$, graded on an abelian group $\mathbb{S}$. We will define in this section products and Steenrod operations on the cohomology of a bisimplicial Θ-coalgebra, and show that they commute with the map in cohomology induced by an augmentation. These products and Steenrod operations will turn out to have special filtration-preserving properties that will be useful when we study the spectral sequence of a bisimplicial Θ-coalgebra in subsequent sections.

We will refer often to the gradings arising from the bisimplicial structure , but have suppressed throughout the chapter the "internal" grading defined by $\mathbb{S}$.

We remind the reader of our conventions for bisimplicial objects, as given in Chapter 1, Section 15. We make further definitions for bisimplicial Θ-modules by analogy with the simplicial case discussed in Chapter 1, Section 9. Thus, if X and Y are bisimplicial Θ-modules, we write $X \times Y$ for the bisimplicial Θ-module defined by $(X \times Y)_{p,q} = X_{p,q} \otimes Y_{p,q}$, with diagonal action of Θ (Definition 1.23), and with vertical and horizontal simplicial operators defined as the tensor products of the corresponding operators on X and Y. $X \times Y$ is called the cross-product of the bisimplicial Θ-modules X and Y. We write $CX \otimes CY$ for the ordinary tensor product of the double Θ-complexes CX and CY:

$$(CX \otimes CY)_{p,q} = \bigoplus_{i+j=p} \bigoplus_{k+l=q} (X_{i,k} \otimes Y_{j,l}), \tag{2.1}$$

with $\partial^v = \partial^v \otimes id + id \otimes \partial^v$ and $\partial^h = \partial^h \otimes id + id \otimes \partial^h$.

We also need a construction $CX \otimes_V CY$ that we call the vertical tensor product of X and Y. This is a mix of the cross-product and tensor product constructions. The vertical tensor product is the double chain complex of Θ-modules defined by:

$$(CX \otimes_V CY)_{p,q} = \bigoplus_{k+l=q} (X_{p,k} \otimes Y_{p,l}). \tag{2.2}$$

Here the horizontal differential is $\partial^h = \sum_i d_i^h \otimes d_i^h$, and the vertical differential is $\partial^v = \partial^v \otimes id + id \otimes \partial^v$.

Now suppose given a cup-k product $\{D_k \mid k \geq 0\}$ and suppose X and Y bisimplicial Θ-modules. We define Θ-linear maps

$$H_r = H_r(X,Y) : C(X \times Y) \to CX \otimes_V CY,$$
$$J_s = J_s(X,Y) : CX \otimes_V CY \to CX \otimes CY,$$

where H_r is homogeneous of bidegree $(0, r)$, and J_s is homogeneous of bidegree $(s, 0)$ as follows. The restriction of H_r to $C(X \times Y)_{p,q}$ is

$$D_r(X_{p,*}\, , Y_{p,*}) : X_{p,q} \otimes Y_{p,q} \to \bigoplus_{i+j=q+r} (X_{p,i} \otimes Y_{p,j}), \tag{2.3}$$

and the restriction of J_s to the summand $X_{p,a} \otimes Y_{p,c}$ of $(CX \otimes_V CY)_{p,q}$ (where $a + c = q$) is

$$D_s(X_{*,a}\, , Y_{*,c}) : X_{p,a} \otimes Y_{p,c} \to \bigoplus_{l+m=p+s} (X_{l,a} \otimes Y_{m,c}). \tag{2.4}$$

Thus, the restriction of J_s to $(CX \otimes_V CY)_{p,*} = C(X_{p,*}) \otimes C(X_{p,*})$ is just the product over all pairs (l,m) of the mappings $J_s^{l,m}$ as defined by (1.151). Clearly:

$$\begin{aligned} \partial^v H_r + H_r \partial^v &= H_{r-1} + T H_{r-1} T \\ \partial^h J_s + J_s \partial^h &= J_{r-1} + T J_{r-1} T \\ \partial^h H_r = H_r \partial^h \quad &\text{and} \quad \partial^v J_s = J_s \partial^v. \end{aligned} \tag{2.5}$$

We now define for each integer $k \geq 0$ and each pair X, Y of bisimplicial Θ-modules a Θ-homomorphism $K_k : C(X \times Y) \to CX \otimes CY$, homogeneous of degree k with respect to total degree, by writing

$$K_k = \sum_{r+s=k} T^r J_s T^r H_r, \tag{2.6}$$

where T^s is the s-fold iterate of the transposition. From the relations (2.5) we find after a short computation that K_0 is a chain map; and that, by analogy with (1.59), we have for each $k \geq 1$

$$\partial K_k + K_k \partial = K_{k-1} + T K_{k-1} T, \tag{2.7}$$

where the ∂'s are total differentials. Thus the family $\{ K_k \mid k \geq 0\}$ can be thought of as a bisimplicial cup-k product. We observe that by analogy with equation (1.60), the K_k can be dualized to give $\mathbb{Z}/2$-linear homomorphisms:

$$K_k^* : \mathrm{Hom}_\Theta(CX \otimes CY, N) \to \mathrm{Hom}_\Theta(C(X \times Y), N) \tag{2.8}$$

for any Θ-module N. These satisfy the duals of the equations (2.7).

It is now apparent how to define cohomology operations on $H^* \mathrm{Hom}_\Theta(CX, N)$ for any bisimplicial Θ-coalgebra X and any commutative Θ-algebra N. We define a map of cochain complexes:

$$\phi : \mathrm{Hom}_\Theta(CX, N) \otimes \mathrm{Hom}_\Theta(CX, N) \to \mathrm{Hom}_\Theta(CX \otimes CX, N) \tag{2.9}$$

by analogy with (1.99) and (1.100). Of course, (1.101) holds in the present context. As in (1.103) the coproduct $\psi : X \to X \times X$ induces a map of cochain complexes $\psi^* : \mathrm{Hom}_\Theta(C(X \times X), N) \to \mathrm{Hom}_\Theta(CX, N)$. (1.104) holds in the present context. We can then define, as in (1.105), a natural map of cochain complexes

$$m : \mathrm{Hom}_\Theta(CX, N) \otimes \mathrm{Hom}_\Theta(CX, N) \to \mathrm{Hom}_\Theta(CX, N) \tag{2.10}$$

by the formula:

$$m = \psi^* K_0^* \phi. \tag{2.11}$$

This cochain operation induces a commutative, associative product:

$$\mu : H^* \mathrm{Hom}_\Theta(CX, N) \otimes H^* \mathrm{Hom}_\Theta(CX, N) \to H^* \mathrm{Hom}_\Theta(CX, N). \tag{2.12}$$

We define further, by analogy with (1.108), a cochain operation

$$S^k : \mathrm{Hom}^n_\Theta(CX, N) \to \mathrm{Hom}^{n+k}_\Theta(CX, N) \tag{2.13}$$

by the customary formula

$$S^k(x) = \psi^* K^*_{n-k} \phi(x \otimes x) + \psi^* K^*_{n-k+1} \phi(x \otimes \delta x). \tag{2.14}$$

The duals of the equations (2.7), together with (1.101) and (1.104) imply by analogy with (1.110) and (1.112) that

$$\delta S^k + S^k \delta = 0, \tag{2.15}$$

and that

(2.16)
$$\Delta^k(x,y) = \psi^*[\delta K^*_{n-k+2}\,\phi(x\otimes \delta y) + \delta K^*_{n-k+1}\,\phi(y\otimes x) + K^*_{n-k+2}\,\phi(\delta x\otimes \delta y)].$$

Here $\Delta^k(x,y) = S^k(x+y) - S^k(x) - S^k(y)$ is the deviation of S^k from additivity, defined as in (1.111) for all $x,y\in \mathrm{Hom}^n_\Theta(CX,N)$. It follows that the cochain operation S^k induces an additive cohomology operation

$$Sq^k : H^n\,\mathrm{Hom}_\Theta(CX,N) \to H^{n+k}\,\mathrm{Hom}_\Theta(CX,N) \tag{2.17}$$

for all $n,k\geq 0$.

Suppose the bisimplicial Θ-coalgebra X equipped with an augmentation $\{\lambda, R\}$, in the sense of Definition 1.123. Then the chain mapping $C\lambda$ of (1.135) induces a mapping of cohomology groups

$$\lambda^* : H^*\,\mathrm{Hom}_\Theta(CR,N) \to H^*\,\mathrm{Hom}_\Theta(CX,N) \tag{2.18}$$

for any Θ-module N. If in addition N is a commutative Θ-algebra then products and Steenrod operations are defined on both sides of (2.18), and we assert:

Proposition 2.1. *Let X be a bisimplicial Θ-coalgebra, let N be a commutative Θ-algebra, and $\{\lambda, R\}$ an augmentation of X. Then the induced map (2.18) commutes with products and with the operations Sq^k.*

Proof. Fix an integer $k\geq 0$ and fix an element $u\in H^n\,\mathrm{Hom}_\Theta(CR,N)$. We will show that:

$$Sq^k\lambda^* u = \lambda^* Sq^k u. \tag{2.19}$$

Let u be represented by a cocycle $x: C_nR\to N$. Then $Sq^k\lambda^*u$ is represented by: $\psi^*K^*_{n-k}\phi(\lambda^*x\otimes\lambda^*x) : C_{n+k}X \to N$. We use (2.3), (2.4), (2.6) and the fact that $J_0 = D_0(X_{*,n};X_{*,n}) : X_{0,n}\otimes X_{0,n} \to X_{0,n}\otimes X_{0,n}$ is the identity. We find that $Sq^k\lambda^*u$ is represented by the composition

$$X_{0,n+k} \xrightarrow{\psi} X_{0,n+k}\otimes X_{0,n+k} \xrightarrow{D^{n,n}_{n-k}(X_{0,*};X_{0,*})} X_{0,n}\otimes X_{0,n} \longrightarrow \tag{2.20}$$
$$\xrightarrow{\lambda\otimes\lambda} R_n\otimes R_n \xrightarrow{x\otimes x} N\otimes N \xrightarrow{\mu} N,$$

where we have written $D^{n,n}_{n-k}$ for the map D_{n-k} followed by projection to the summand $X_{0,n}\otimes X_{0,n}\subset (CX\otimes CX)_{0,2n}$. But $D^{n,n}_{n-k}$ is natural with respect to the map of simplicial objects $\lambda: X_{0,*}\to R_*$, and since λ preserves the coproduct we can write (2.20) in the form $\mu\circ(x\otimes x)\circ D^{n,n}_{n-k}(R,R)\circ\psi\circ\lambda : X_{0,n-k}\to N$. But this cochain represents λ^*Sq^ku, so (2.19) is proved. A similar argument shows that λ^* preserves products. □

Remark 2.2. The products (2.12) and the squaring operations (2.17) on the cohomology of bisimplicial coalgebras can be shown independent of choice of the cup-k product $\{D_k \mid k\geq 0\}$; and they can be shown to satisfy the Cartan formula and the Adem relations.

The proofs of the statements in Remark (2.2) are not at all difficult, but we will not write them out. The reason is that in all the cases in which we are interested, the bisimplicial coalgebra X will come equipped with an augmentation to a simplicial coalgebra, for which the induced map in cohomology (2.18) is an isomorphism commuting with both products and Steenrod squares. Thus, the statements above about the cohomology of bisimplicial coalgebras will follow from Proposition 1.111: the corresponding statements in the simplicial case.

2. Filtrations

Fix a bialgebra Θ over $\mathbb{Z}/2$, graded on an abelian group $\mathbb{S}$. Fix a bisimplicial Θ-coalgebra X and a commutative Θ-algebra N. In Chapter 1, Section 16 we recalled the standard filtrations on the total chains CX, the cochains $\mathrm{Hom}_\Theta(CX, N)$, and on the cohomology groups $H^* \mathrm{Hom}_\Theta(CX, N)$. Here we study how the cochain and cohomology operations defined in the previous section behave with respect to these filtrations. The main results are Proposition 2.6, which tells how the cup-k products treat filtration; and Proposition 2.9, which tells how the operations Sq^k treat filtration. We will then use the latter result to define squaring operations on the "associated graded" module $E_0 H^* \mathrm{Hom}_\Theta(CX, N)$ as it has been defined by (1.140).

We begin with:

REMARK 2.3. Let X be a bisimplicial Θ-coalgebra. Since $\psi : X \to X \times X$ is a map of bisimplicial Θ-modules, the induced map

$$\psi^* : H^* \mathrm{Hom}_\Theta(C(X \times X), N) \to H^* \mathrm{Hom}_\Theta(CX, N) \tag{2.21}$$

preserves filtration.

To go further we have to define filtrations on various tensor products. Suppose X and Y are both bisimplicial Θ-modules. Then $CX \otimes CY$ is a tensor product of filtered complexes, and so is equipped with the tensor product filtration:

$$F_p(CX \otimes CY) = \sum_{i+j=p} F_i CX \otimes F_j CX. \tag{2.22}$$

Also the dual cochain complex $\mathrm{Hom}_\Theta(CX \otimes CY, N)$ has the descending filtration dual to (2.22), defined by analogy with (1.138). Finally, the tensor product of cochain complexes $\mathrm{Hom}_\Theta(CX, N) \otimes \mathrm{Hom}_\Theta(CX, N)$ has the tensor product filtration:

$$\begin{aligned} F^p[\mathrm{Hom}_\Theta(CX, N) \otimes \mathrm{Hom}_\Theta(CX, N)] \\ = \sum_{i+j=p} F^i \mathrm{Hom}_\Theta(CX, N) \otimes F^j \mathrm{Hom}_\Theta(CX, N). \end{aligned} \tag{2.23}$$

REMARK 2.4. With these definitions, ϕ of (2.9) preserves filtration.

Now we study how the cup-k products treat filtration.

PROPOSITION 2.5. *Suppose the cup-k product $\{D_k \mid k \geq 0\}$ is "special" in the sense of Definition 1.94. Then the elements of the corresponding bisimplicial cup-k product $\{K_k \mid k \geq 0\}$, as in* (2.6)*, satisfy both $K_k F_p C(X \times Y) \subseteq F_{p+k}(CX \otimes CY)$ and $K_k F_p C(X \times Y) \subseteq F_{2p}(CX \otimes CY)$ for all pairs X, Y of bisimplicial Θ-modules.*

PROOF. The first statement follows from the definition (2.6) of K_k, and the second from (2.3), (2.4), (2.6), and the first assertion of Proposition 1.96. □

We remind the reader of the conventions concerning descending filtrations that we gave in Remark 1.125. As an immediate consequence of the above proposition we have:

PROPOSITION 2.6. *Let X,Y be any pair of bisimplicial Θ-modules; and suppose given a cup-k product $\{D_k \mid k \geq 0\}$ that is special. Then the corresponding homomorphisms K_k^* of (2.8) satisfy both*

$$K_k^*(F^p) \subseteq F^{\lceil p/2 \rceil} \tag{2.24a}$$

$$K_k^*(F^p) \subseteq F^{p-k} \tag{2.24b}$$

for all $k \geq 0$ and all integers p. (Here by $\lceil n \rceil$ we mean the least integer greater than or equal to n.)

Now we can study the effect of multiplication and squaring operations on filtration. We begin with:

PROPOSITION 2.7. *Let X be a bisimplicial Θ-coalgebra, and N a commutative Θ-algebra. Then the cochain operation m of (2.10) preserves filtration.*

PROOF. This follows at once from equations (2.11) and (2.24), and Remarks 2.3 and 2.4. □

Recall $E_0^{p,q}H^*$ from (1.140). As a consequence of the last proposition, the multiplication μ of (2.12) defines a pairing $E_0^{p,q} \otimes E_0^{r,s} \to E_0^{p+r,q+s}$ for all $p, q, r, s \geq 0$, and so defines a homomorphism of bigraded vector spaces:

$$E_0(\mu) : E_0H^* \otimes E_0H^* \to E_0H^*.$$

In a similar spirit:

PROPOSITION 2.8. *Let X be a bisimplicial Θ-coalgebra, and N a commutative Θ-algebra. Then the cochain operation S^k of (2.13) preserves filtration.*

PROOF. This follows at once from equations (2.14) and (2.24a), and Remarks 2.3 and 2.4. □

We would like next to say in what sense Sq^k of (2.17) passes to a homomorphism of the bigraded vector space E_0H^*. Surely the result just proved implies that:

$$Sq^k\left(F^pH^{p+q}\operatorname{Hom}_\Theta(CX,N)\right) \subseteq F^pH^{p+q+k}\operatorname{Hom}_\Theta(CX,N) \tag{2.25}$$

and so Sq^k would define a homomorphism:

$$E_0(Sq^k) : E_0^{p,q}H^* \to E_0^{p,q+k}H^*. \tag{2.26}$$

But it turns out that (2.25) is sharp only when $k \leq q$. When $k > q$, the operator Sq^k sends F^pH^{p+q} to a higher filtration, and the homomorphism (2.26) is zero. The following proposition demonstrates this, and also points the way to the correct definition of $E_0(Sq^k)$.

PROPOSITION 2.9. *Let X be a bisimplicial Θ-coalgebra, and N a commutative Θ-algebra. Then for all $k, p, q \geq 0$ we have both:*

$$Sq^k\left(F^pH^{p+q}\operatorname{Hom}_\Theta(CX,N)\right) \subseteq F^pH^{p+q+k}\operatorname{Hom}_\Theta(CX,N) \tag{2.27a}$$

$$Sq^k\left(F^pH^{p+q}\operatorname{Hom}_\Theta(CX,N)\right) \subseteq F^{p+k-q}H^{p+q+k}\operatorname{Hom}_\Theta(CX,N). \tag{2.27b}$$

PROOF. Let α be any element of $F^pH^{p+q}\operatorname{Hom}_\Theta(CX,N)$ and suppose α is represented by a cocyle $x \in F^p(\operatorname{Hom}_\Theta^{p+q}(CX,N))$. Then $Sq^k(x)$ is represented by $S^k(x)$ as given by (2.14). Since $\delta(x) = 0$ this is $S^k(x) = \psi^* K^*_{p+q-k}\phi(x \otimes x)$. Remarks 2.3 and 2.4 and equation (2.24a) show that $S^k(x) \in F^p\operatorname{Hom}_\Theta(CX,N)$. Remarks 2.3 and 2.4 and equation (2.24b) show that $S^k(x) \in F^{p+k-q}\operatorname{Hom}_\Theta(CX,N)$. □

Of course (2.27b) is interesting only in those cases when it is stronger than (2.27a): when $k > q$.

Let X be a bisimplicial Θ-coalgebra, and N a commutative Θ-algebra. With E_0H^* defined by (1.140) we define for each $k \geq 0$ a $\mathbb{Z}/2$-linear homomorphism

(2.28) $$E_0(Sq^k):\ E_0H^* \to E_0H^*$$

by writing:

(2.28a) $$E_0(Sq^k): E_0^{p,q}H^* \to E_0^{p,q+k}H^* \qquad \text{if } 0 \leq k \leq q$$

(2.28b) $$E_0(Sq^k): E_0^{p,q}H^* \to E_0^{p+k-q,2q}H^* \qquad \text{if } q \leq k,$$

where the first of these is induced by Sq^k of (2.27a) and the second by Sq^k of (2.27b). Equations (2.27a) and (2.27b) also assure that our $E_0(Sq^k)$ is well defined.

3. The Spectral Sequence

In this section we use the results of the previous one to study products and Steenrod operations in the spectral sequence of a bisimplicial coalgebra. The first main result is Theorem 2.15, which shows how to get well defined Steenrod operations with domains the E_r page of the spectral sequence, for each $r \geq 2$. Theorem 2.16 then shows that the operations on E_2 determine the operations on E_r for $r \geq 2$. Theorem 2.17 describes how the Steenrod operations commute with the differentials of the spectral sequence. Theorem 2.21 is the survival result for the cup-square that was advertised at the beginning of this chapter. Theorem 2.22 shows that the Steenrod operations on E_∞ are compatible with those we defined on E_0H^* in the previous section. Theorem 2.23 gives a general description of the Steenrod operations on E_2 as certain operations on cochains. Here we divide the operations on E_2 into "vertical" and "diagonal", depending on how they move elements on the $P-Q$ plane of the spectral sequence. This description of the operations at the E_2-level will be useful when we calculate the operations on the E_2 terms of the change-of-rings and Eilenberg-Moore spectral sequences, in Chapters 5 and 7, respectively.

Throughout we fix a bialgebra Θ over $\mathbb{Z}/2$, graded on an abelian group $\mathbb{S}$. We fix a bisimplicial Θ-coalgebra X and a commutative Θ-algebra N. In Chapter 1, Section 16 we reviewed "the spectral sequence of X with coefficients in N": this is just the spectral sequence associated with the standard filtration on the cochains $\operatorname{Hom}_\Theta(CX,N)$. We defined the subspaces $B_r^{p,q} \subseteq Z_r^{p,q} \subseteq F^p\operatorname{Hom}_\Theta^{p+q}(CX,N)$, and the terms E_r of the spectral sequence. Now we study the effect of the cochain operations m of (2.10) and S^k of (2.13) on these vector spaces, and their relationship to the differentials of the spectral sequence.

PROPOSITION 2.10. *For all $l, p \geq 0$ and $r \geq 1$ we have*

$$m(Z_r^{l,*} \otimes Z_r^{p,*}) \subseteq Z_r^{l+p,*}. \tag{2.29}$$

Further, m passes to a well-defined homomorphism

$$\mu : E_r^{l,*} \otimes E_r^{p,*} \to E_r^{l+p,*} \tag{2.30}$$

satisfying for all $\alpha \in E_r^{l,*}$ and $\beta \in E_r^{p,*}$:

$$d_r\mu(\alpha \otimes \beta) = \mu(d_r\alpha \otimes \beta + \alpha \otimes d_r\beta). \tag{2.31}$$

PROOF. Equation (2.29) follows immediately from Proposition 2.7 and the fact that m is a map of cochain complexes. To show that (2.30) is well-defined we must show both:

$$m(B_r^{l,*} \otimes Z_r^{p,*}) \subseteq B_r^{l+p,*} \tag{2.32a}$$

$$m(Z_r^{l,*} \otimes B_r^{p,*}) \subseteq B_r^{l+p,*}. \tag{2.32b}$$

It suffices to show just the first of these. This task, in turn, breaks up into showing both that

$$m(\delta Z_{r-1}^{l-r+1,*} \otimes Z_r^{p,*}) \subseteq B_r^{l+p,*} \tag{2.33}$$

and that:

$$m(Z_{r-1}^{l+1,*} \otimes Z_r^{p,*}) \subseteq B_r^{l+p,*}. \tag{2.34}$$

To show the first of these, suppose $x \in Z_{r-1}^{l-r+1,*}$ and $y \in Z_r^{p,*}$. Then $m(\delta x \otimes y) = m(x \otimes \delta y) + \delta m(x \otimes y)$. Using Proposition 2.7 and the fact that m commutes with the differential, one sees that $m(x \otimes \delta y) \in Z_{r-1}^{l+p+1,*}$, and that $\delta m(x \otimes y) \in \delta Z_{r-1}^{l+p-r+1,*}$. So (2.33) is proved. Then to show (2.34) it will be enough to show that $m(Z_{r-1}^{l+1,*} \otimes Z_r^{p,*}) \subseteq Z_{r-1}^{l+p+1,*}$. But this also follows from Proposition 2.7 and the fact that m commutes with the differential. This completes the construction of the homomorphism (2.30). The fact that m commutes with the differential also implies (2.31). □

Choosing r sufficiently large in (2.30) we obtain for all $l, p \geq 0$ a homomorphism:

$$\mu : E_\infty^{l,*} \otimes E_\infty^{p,*} \to E_\infty^{l+p,*}. \tag{2.35}$$

This product on E_∞ is compatible with the product on E_0H^* we have previously defined. With $\pi : E_\infty \to E_0H^*$ the isomorphism of (1.148) we have:

PROPOSITION 2.11. *The following diagram commutes:*

$$\begin{array}{ccc} E_\infty \otimes E_\infty & \xrightarrow{\mu} & E_\infty \\ {\scriptstyle \pi\otimes\pi}\downarrow & & \downarrow{\scriptstyle \pi} \\ E_0H^* \otimes E_0H^* & \xrightarrow{E_0(\mu)} & E_0H^*. \end{array}$$

PROOF. Both μ and $E_0(\mu)$ are induced by the cochain operation m of (2.11). □

Having studied the effect of m on the terms of the spectral sequence, we turn to the operators S^k. We would like to define squaring operations from each E_r term of the spectral sequence. In view of our definitions (2.28) and the isomorphism (1.148), it seems clear that these operations should affect the various gradings according to the scheme:

$$Sq^k : E_r^{p,q} \to E_t^{p,q+k} \qquad \text{if} \quad 0 \le k \le q, \tag{2.36a}$$

$$Sq^k : E_r^{p,q} \to E_t^{p+k-q,2q} \qquad \text{if} \quad q \le k. \tag{2.36b}$$

Ideally we would like to be able to choose $t = r$, so that we could say that E_r is a module over the Steenrod algebra. For the operations (2.36a) we will indeed be able to do this. In view of Proposition 2.10 one might think one could take $t = r$ in (2.36b) as well. After all, Sq^{p+q} operating on $E_r^{p,q}$ should be the cup-square, and we have just shown that products are well-defined on E_r. But the obstacle to choosing $t = r$ in (2.36b) lies in establishing the analogues of (2.32) and (2.33). In order to get a well-defined operation $Sq^k : E_r^{p,q} \to E_r^{p+k-q,2q}$ for $k > q$ one would have to show that $S^k(B_r^{p,q}) \subseteq B_r^{p+k-q,2q}$, and in particular that $S^k(\delta Z_{r-1}^{p-r+1,q+r-2}) \subseteq B_r^{p+k-q,2q}$. So one would have to show that if $x \in Z_{r-1}^{p-r+1,q+r-2}$ then:

$$S^k(\delta x) = \psi^* \, K_{p+q-k}^* \, \phi\,(\delta x \otimes \delta x) \, \in \, B_r^{p+k-q,2q}. \tag{2.37}$$

It is easy to prove (2.37) in the case $k = p+q$. One use the fact that K_0^* is a filtration-preserving chain map, and the equation $\phi(\delta x \otimes \delta x) = \delta\phi(x \otimes \delta x)$, to show immediately that $\psi^* \, K_0^* \, \phi\,(\delta x \otimes \delta x) \in \delta Z_{r-1}^{2p-r+1,2q+r-2} \subseteq B_r^{2p,2q}$. But this argument breaks down if $k < p+q$, and we are left with the necessity of choosing in (2.36b) some value of t that is greater than r.

The following three propositions lay the groundwork for defining the operations (2.36a) and (2.36b), and for choosing the right value of t.

PROPOSITION 2.12. *For all integers p, q, r and all integers $k \ge 0$ we have:*

$$S^k(Z_r^{p,q}) \subseteq Z_r^{p,q+k}, \tag{2.38a}$$

$$S^k(Z_r^{p,q}) \subseteq Z_{2r+k-q-1}^{p,q+k}, \tag{2.38b}$$

$$S^k(Z_r^{p,q}) \subseteq Z_t^{p+k-q,2q} \text{ for all } t \text{ with } \; 0 \le t \le 2r-1. \tag{2.38c}$$

PROOF. Suppose that $x \in F^p \operatorname{Hom}_\Theta^{p+q}(CX, N)$ and $\delta x \in F^{p+r} \operatorname{Hom}_\Theta(CX, N)$. Then it follows immediately from Proposition 2.8 and (2.15) that $S^k x \in F^p$ and $\delta S^k x \in F^{p+r}$. So (2.38a) is established. To establish (2.38b), suppose again that $x \in F^p \operatorname{Hom}_\Theta^{p+q}(CX, N)$ and $\delta x \in F^{p+r}$. We need only show $\delta S^k x \in F^{p+2r+k-q-1}$. But $\delta S^k x = S^k \delta x = \psi^* K_{p+q+1-k}^* \phi(\delta x \otimes \delta x)$. Appeal to Remarks 2.3 and 2.4 and (2.24b) shows that we have $\delta S^k x \in F^{p+2r+k-q-1}$ as required. To establish (2.38c) in general, it is enough to prove it in the case $t = 2r-1$, and appeal to (1.142). So suppose again that $x \in F^p \operatorname{Hom}_\Theta^{p+q}(CX, N)$ and $\delta x \in F^{p+r} \operatorname{Hom}_\Theta(CX, N)$. Then by Remarks 2.3 and 2.4 and equations (2.24b) and (2.14) we have $S^k x \in F^{p+k-q}$, and $S^k \delta x = \psi^* K_{p+q+1-k}^* \phi(\delta x \otimes \delta x) \in F^{p+k-q+2r-1}$. Hence $\delta S^k x = S^k \delta x$ in $F^{p+2r+k-q-1}$, so (2.38c) is proved. □

Note that equations (2.38) are meaningful and valid even when some of the subscripts and superscripts are negative; e.g., when $p+k < q$ or when $k < q-2r+1$.

Since the cochain operation S^k is not additive, we must next determine how the deviation from additivity, Δ^k of (2.16), acts on the spaces $Z_r^{p,q}$.

PROPOSITION 2.13. *Suppose that $p, q, k \geq 0$ and $r \geq 2$, and suppose that x and y are in $Z_r^{p,q}$. Then*

(2.39a) $$\Delta^k(x,y) \in B_r^{p,q+k},$$

(2.39b) $$\Delta^k(x,y) \in B_{r+k-q}^{p+k-q,2q} \quad \text{if} \quad k \leq q+r-2,$$

(2.39c) $$\Delta^k(x,y) \in B_{2r-2}^{p+k-q,2q}.$$

PROOF. We begin with (2.39a) and use (2.16). It is enough to show that with $n = p+q$ both $\psi^* K^*_{n-k+2}\,\phi(x \otimes \delta y)$ and $\psi^* K^*_{n-k+1}\,\phi(y \otimes x)$ lie in $Z_{r-1}^{p-r+1,*}$, and that $\psi^* K^*_{n-k+2}\,\phi(\delta x \otimes \delta y) \in Z_{r-1}^{p+1,*}$. But all these statements follow easily from the assumptions $x, y \in F^p$, δx, $\delta y \in F^{p+r}$, from equation (2.24a) and from Remarks 2.3 and 2.4. To show (2.39b) we again appeal to (2.16). It is enough to show that with $n = p+q$ and $k \leq q+r-2$ we have all of:

(2.40a) $$\psi^* K^*_{n-k+2}\,\phi(x \otimes \delta y) \in Z_{r+k-q-1}^{p-r+1,*},$$

(2.40b) $$\psi^* K^*_{n-k+1}\,\phi(y \otimes x) \in Z_{r+k-q-1}^{p-r+1,*},$$

(2.40c) $$\psi^* K^*_{n-k+2}\phi(\delta x \otimes \delta y) \in Z_{r+k-q-1}^{p+k-q+1,*}.$$

To show (2.40a), we need to show both $\psi^* K^*_{n-k+2}\,\phi(x \otimes \delta y) \in F^{p-r+1}$ and that $\delta\psi^* K^*_{n-k+2}\,\phi(x \otimes \delta y) \in F^{p+k-q}$. But $\phi(x \otimes \delta y)$ lies in F^{2p+r}, so the first statement follows from (2.24a), and the second from (2.24b). To show (2.40b) we need both $\psi^* K^*_{n-k+1}\,\phi(y \otimes x) \in F^{p-r+1}$ and that $\delta\psi^* K^*_{n-k+1}\,\phi(y \otimes x) \in F^{p+k-q}$. But $\phi(y \otimes x)$ lies in F^{2p}, so the first statement follows from (2.24a). The second requires a little more work: we use the dual of (2.7) with (1.101) and (1.104) to write:

(2.41) $$\begin{aligned}\delta\psi^* K^*_{n-k+1}\,\phi(y \otimes x) &= \psi^* K^*_{n-k+1}\,\phi[(\delta y \otimes x) + (y \otimes \delta x)] \\ &\quad + \psi^* K^*_{n-k}\phi[(y \otimes x) + (x \otimes y)].\end{aligned}$$

But $\delta y \otimes x$ and $y \otimes \delta x$ both lie in F^{2p+r}, and $y \otimes x$ and $x \otimes y$ both lie in F^{2p}, so (2.24b) shows that all terms on the right hand side of (2.41) lie in F^{p+k-q}. This completes the proof of (2.40b). To show (2.40c) we need to show that both $\psi^* K^*_{n-k+2}\,\phi(\delta y \otimes \delta x) \in F^{p+k-q+1}$ and that $\delta\psi^* K^*_{n-k+2}\,\phi(\delta y \otimes \delta x) \in F^{p+r+2k-2q}$. But $\delta y \otimes \delta x$ lies in F^{2p+2r} so the first statement follows from (2.24b). The second follows from (2.24b) in conjunction with the hypothesis $k \leq q+r-2$ that goes with (2.39b). These arguments complete the proof of (2.39b). To show (2.39c) we use a similar argument. By (2.16), it is enough to show that with $n = p+q$ we have all of:

(2.42a) $$\psi^* K^*_{n-k+2}\,\phi(x \otimes \delta y) \in Z_{2r-3}^{p+k-q-2r+3,*},$$

(2.42b) $$\psi^* K^*_{n-k+1}\,\phi(y \otimes x) \in Z_{2r-3}^{p+k-q-2r+3,*},$$

(2.42c) $$\psi^* K^*_{n-k+2}\phi(\delta x \otimes \delta y) \in Z_{2r-3}^{p+k-q+1,*}.$$

Since $\phi(x \otimes \delta y) \in F^{2p+r}$, (2.42a) follows easily from (2.24b) and the assumption $r \geq 2$. To show (2.42b) we need both $\psi^* K^*_{n-k+1}\,\phi(y \otimes x) \in F^{p+k-q-2r+3}$ and that $\delta\psi^* K^*_{n-k+1}\,\phi(y \otimes x) \in F^{p+k-q}$. Since $\phi(y \otimes x)$ lies in F^{2p}, the first of these follows from (2.24b) and the assumption $r \geq 2$. The proof of the second statement is based on (2.41), and is word-for-word identical to the argument that follows that equation. So (2.42b) is shown. To show (2.42c) we need both $\psi^* K^*_{n-k+2}\phi(\delta x \otimes \delta y) \in F^{p+k-q+1}$ and that $\delta\psi^* K^*_{n-k+2}\phi(\delta x \otimes \delta y) \in F^{p+k-q+2r-2}$. But $\phi(\delta x \otimes \delta y)$ lies in F^{2p+2r}, so both statements follow at once from Remarks 2.3 and 2.4 and equation (2.24b). So (2.42c) is shown. This completes the proof of (2.39c), and of Proposition 2.13. □

One more result is needed to define the operations (2.36).

PROPOSITION 2.14. *For all $p, q, k \geq 0$ and all $r \geq 2$ we have the following inclusions:*

(2.43a) $$S^k(B_r^{p,q}) \subseteq B_r^{p,q+k},$$

(2.43b) $$S^k(B_r^{p,q}) \subseteq B_{r+k-q}^{p+k-q,2q} \text{ if } \quad k \leq q+r-2,$$

(2.43c) $$S^k(B_r^{p,q}) \subseteq B_{2r-2}^{p+k-q,2q}.$$

PROOF. Proposition 2.13 implies that, in order to show (2.43a), it suffices to show that both $S^k(\delta Z_{r-1}^{p-r+1,*}) \subseteq \delta Z_{r-1}^{p-r+1,*}$, and that $S^k(Z_{r-1}^{p+1,*}) \subseteq Z_{r-1}^{p+1,*}$. But these statements are immediate consequences of (2.15) and (2.38a). To show (2.43b) we need both $S^k(\delta Z_{r-1}^{p-r+1,q+r-2}) \subseteq \delta Z_{r+k-q-1}^{p-r+1,*}$ and $S^k(Z_{r-1}^{p+1,q-1}) \subseteq Z_{r+k-q-1}^{p+k-q+1,*}$. (Of course, Proposition 2.13 is involved in this reduction also.) The first of these follows from (2.15) and (2.38b). To show the second, note the hypothesis $k \leq q+r-2$ allows us to deduce from (2.38c) that $S^k(Z_{r-1}^{p+1,q-1}) \subseteq Z_{r+k-q-2}^{p+k-q+2,*}$. But by (1.142) this latter group is contained in $Z_{r+k-q-1}^{p+k-q+1,*}$. This completes the proof of (2.43b). To show (2.43c) we must show both

(2.44) $$S^k(\delta Z_{r-1}^{p-r+1,q+r-2}) \subseteq \delta Z_{2r-3}^{p+k-q-2r+3,*}$$

and

(2.45) $$S^k(Z_{r-1}^{p+1,q-1}) \subseteq Z_{2r-3}^{p+k-q+1,*}.$$

But we may apply (2.38c) to conclude that $S^k(Z_{r-1}^{p-r+1,q+r-2}) \subseteq Z_{2r-3}^{p+k-q-2r+3,*}$, so (2.44) follows from (2.15). To show that (2.45) holds, note that (2.38c) implies that $S^k(Z_{r-1}^{p+1,q-1}) \subseteq Z_{2r-3}^{p+k-q+2,*}$, and use (1.142). □

Now we are ready to define Steenrod squaring operations from the E_r-term of the spectral sequence.

THEOREM 2.15. *For all $p, q \geq 0$ and all $r \geq 2$, the cochain operations S^k of* (2.13) *pass to well-defined vector space homomorphisms:*

(2.46a) $$Sq^k : E_r^{p,q} \to E_r^{p,q+k} \quad \text{if} \quad 0 \leq k \leq q,$$

(2.46b) $$Sq^k : E_r^{p,q} \to E_{r+k-q}^{p+k-q,2q} \quad \text{if} \quad q \leq k \leq q+r-2,$$

(2.46c) $$Sq^k : E_r^{p,q} \to E_{2r-2}^{p+k-q,2q} \quad \text{if} \quad q+r-2 \leq k.$$

PROOF. This is now an immediate consequence of Propositions 2.12, 2.13, and 2.14. □

Note in particular that Sq^k is defined as an operator from E_2 to E_2, for all values of $k \geq 0$.

In fact, the operations at the E_2 level determine the operations on E_r for all $r \geq 2$. That is the content of the next theorem. In what follows, if α is an element in the E_s-term of the spectral sequence that survives to the E_r-term for some $r \geq s$, we will write $[\alpha]_r$ for the class in E_r that α represents.

THEOREM 2.16. *Let $p, q, k \geq 0$ and $r \geq 2$ be given, and let $\alpha \in E_2^{p,q}$ be given. Suppose α survives to represent a class $[\alpha]_r \in E_r^{p,q}$. Then $Sq^k\alpha \in E_2$ survives to represent a class $[Sq^k\alpha]_t \in E_t$, where*

$$t = \begin{cases} r, & \text{if} \quad 0 \leq k \leq q; \\ r + k - q, & \text{if} \quad q \leq k \leq q + r - 2; \\ 2r - 2, & \text{if} \quad q + r - 2 \leq k. \end{cases}$$

Further, we have in all cases:

$$Sq^k [\alpha]_r = [Sq^k\alpha]_t . \tag{2.47}$$

PROOF. The proofs in all three cases are similar; we offer as an example the case $q \leq k \leq q+r-2$. Suppose $\alpha \in E_2^{p,q}$ is represented by a cochain $x \in Z_2^{p,q}$. Then $Sq^k\alpha \in E_2^{p+k-q,*}$ is represented by $S^kx \in Z_2^{p+k-q,*}$. To show that $Sq^k\alpha$ survives to $E_{r+k-q}^{p+k-q,*}$ we must show that $\delta S^kx \in F^{p+r+2k-2q}$. But since α survives to $E_r^{p,q}$ we have $\delta x \in F^{p+r}$. So (2.38c) implies that $S^k\delta x \in F^{p+2r+k-q-1}$. So by (2.15), and the assumption that $k \leq q + r - 2$, we have that $\delta S^kx = S^k\delta x \in F^{p+2r+k-q-1} \subseteq F^{p+r+2k-2q}$, as required. Equation (2.47) is an immediate consequence of the definitions: both sides are represented by the cochain S^kx. □

The next theorem shows how the squaring operations commute with the differentials of the spectral sequence.

THEOREM 2.17. *Let integers $r \geq 2$ and $p, q, k \geq 0$ be given, and let $\alpha \in E_r^{p,q}$ be given. Then both $Sq^k\alpha$ and $Sq^kd_r\alpha$ survive to E_t, where*

$$t = \begin{cases} r, & \text{if} \quad 0 \leq k \leq q - r + 1; \\ 2r + k - q - 1, & \text{if} \quad q - r + 1 \leq k \leq q; \\ 2r - 1, & \text{if} \quad q \leq k. \end{cases} \tag{2.48}$$

Further, in E_t we have the relation:

$$d_t [Sq^k\alpha]_t = [Sq^k d_r \alpha]_t . \tag{2.49}$$

PROOF. All of these statements are consequences of the definitions in Theorem 2.15, and of equation (2.15). We write out the proofs in some cases. Consider for example the case $q-r+1 \leq k \leq q$. Let $\alpha \in E_r^{p,q}$ be given, and suppose α represented by an element $x \in Z_r^{p,q}$. Then $Sq^k\alpha \in E_r^{p,q+k}$ is represented by $S^kx \in Z_r^{p,q+k}$, where we are using (2.38a). To show that $Sq^k\alpha$ survives to $E_{2r+k-q-1}^{p,*}$, we must show that $\delta S^kx \in F^{p+k-q+2r-1}$. But $\delta x \in F^{p+r}$, so $\delta x \in Z_t^{p+r,q-r+1}$ for all values of $t \geq 0$. Then (2.38c) implies that $S^k\delta x \in F^{p+k-q+2r-1}$, so $\delta S^kx \in F^{p+k-q+2r-1}$ as required. Also, (2.49) follows at once from (2.15). Consider next the case $q \leq k$. Let $\alpha \in E_r^{p,q}$ be given, and suppose α represented by an element $x \in Z_r^{p,q}$. Consider first the subcase $q \leq k \leq q+r-2$. Then $Sq^k\alpha \in E_{r+k-q}^{p+k-q,*}$ is represented by $S^kx \in Z_{r+k-q}^{p+k-q,*}$; here we are using (2.38c). To show that $Sq^k\alpha$ survives to $E_{2r-1}^{p+k-q,*}$, we must show that $\delta S^kx \in F^{p+k-q+2r-1}$. But $\delta S^kx = S^k\delta x$, and since $\delta x \in F^{p+r}$, so $\delta x \in Z_t^{p+r,q-r+1}$ for all values of $t \geq 0$. Hence (2.38c) implies that $S^k\delta x \in F^{p+k-q+2r-1}$, as required. Also, (2.49) follows at once from (2.15). The subcase $k > q + r - 2$ is similar. Here $Sq^k\alpha \in E_{2r-2}^{p+k-q,*}$ is represented by

$S^k x \in Z_{2r-2}^{p+k-q,*}$, where we are using (2.38c). The proof that $Sq^k\alpha$ survives to $E_{2r-1}^{p+k-q,*}$ proceeds just as in the previous subcase, and the relation (2.49) follows from (2.15), as before. □

It is interesting to draw diagrams of the relation (2.49) in the $P-Q$ plane. The four cases are: $0 \le k \le q-r+1$; $q-r+1 < k \le q-1$; $q \le k \le q+r-2$; $q+r-2 < k$. The diagrams for these situations differ somewhat from one another. We show here the case $q \le k \le q+r-2$. The reader is invited to draw some of the others.

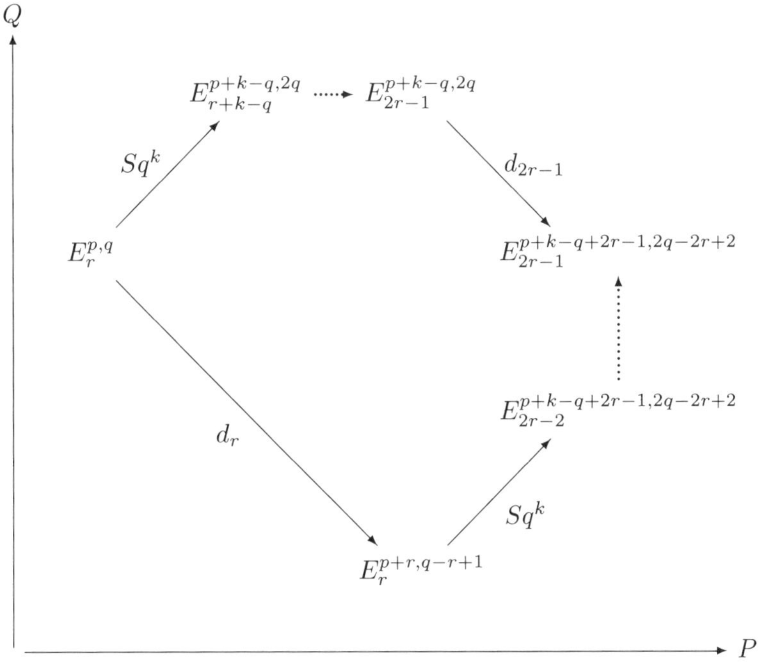

The case $\mathbf{q \le k \le q+r-2}$

In this drawing, the dotted arrows do not indicate any displacement in the $P-Q$ plane. Rather, they are meant to show survival of an element from one page of the spectral sequence to a later one.

Recall that for each integer $n \ge 2$ the differential $d_n : E_n^{0,n-1} \to E_n^{n,0}$ is called the transgression τ_n. A special case of Theorem 2.17 asserts that transgression commutes with Steenrod operations. For the special case of the Serre spectral sequence, this is known as the Kudo Transgression Theorem [**37**].

COROLLARY 2.18. *Let integers k, n be given with $n \ge 2$ and $0 \le k \le n-1$. Let $\tau_n : E_n^{0,n-1} \to E_n^{n,0}$ be the transgression. Then for each $\alpha \in E_n^{0,n-1}$ both $Sq^k\alpha$ and $Sq^k\tau_n\alpha$ survive to E_{n+k}, and*

$$[Sq^k\tau_n\alpha]_{n+k} = \tau_{n+k}\,[Sq^k\alpha]_{n+k}\,. \tag{2.50}$$

PROOF. (2.50) is just a special case of (2.49), with the value of t being given by the second of equations (2.48). □

Here is another useful (and immediate) corollary of Theorem 2.17:

COROLLARY 2.19. *Let $\alpha \in E_2^{p,q}$ be an infinite cycle. Then so is $Sq^k\alpha$, for all $k \geq 0$.*

Classically, in algebras over the Steenrod algebra, Sq^n of an n-dimensional element is equal to the cup-square of that element. The following theorem is a version of this result that applies when Steenrod operations are defined on the terms of a spectral sequence.

THEOREM 2.20. *Suppose $r \geq 2$ and let $\alpha \in E_r^{p,q}$ be given. Then the cup-square $\alpha^2 \in E_r^{2p,2q}$ survives to E_t, where*

$$t = \begin{cases} r+p, & \text{if } \ p \leq r-2 \\ 2r-2 & \text{if } \ r-2 \leq p, \end{cases} \tag{2.51}$$

and

$$[\alpha^2]_t = Sq^{p+q}\alpha. \tag{2.52}$$

PROOF. Let $\alpha \in E_r^{p,q}$ be represented by a cochain $x \in Z_r^{p,q}$. It is enough to show that $\alpha^2 \in E_r^{2p,2q}$ can be represented by the cochain $S^{p+q}x \in Z_r^{2p,2q}$. To show this, it suffices to show that the cochain $\psi^* K_1^* \phi(x \otimes \delta x)$ represents zero in $E_r^{2p,2q}$; so in particular it suffices to show that:

$$\psi^* K_1^* \phi(x \otimes \delta x) \in Z_{r-1}^{2p+1,2q-1}. \tag{2.53}$$

But $\delta x \in F^{p+r}$, so from (2.24b) and from Remarks 2.3, 2.4 it follows that the element $\psi^* K_1^* \phi(x \otimes \delta x)$ lies in F^{2p+1}. So to complete the proof of (2.53) we need only show that $\delta\psi^* K_1^* \phi(x \otimes \delta x) \in F^{2p+r}$. But from the dual of (2.7) we obtain

$$\delta\psi^* K_1^* \phi(x \otimes \delta x) = \psi^* K_1^* \phi(\delta x \otimes \delta x) + \psi^* K_0^* \phi(x \otimes \delta x) + \psi^* K_0^* \phi(\delta x \otimes x), \tag{2.54}$$

and each term on the right-hand side of (2.54) clearly lies in F^{2p+r}. So (2.53) is proved, and we are done. □

We have shown that if $\alpha \in E_r^{p,*}$ then α^2 survives to E_t, with t given by (2.51). This is the value of t needed in order that (2.52) make sense. But α^2 actually survives longer than that.

THEOREM 2.21. *Suppose $r \geq 2$ and let $\alpha \in E_r$ be given. Then α^2 survives to E_{2r-1}.*

PROOF. In the proof of the previous theorem we showed that if $\alpha \in E_r^{p,q}$ is represented by a cochain $x \in Z_r^{p,q}$ then $\alpha^2 \in E_r^{2p,2q}$ can be represented by the cochain $S^{p+q}x \in Z_r^{2p,2q}$. So it will suffice to show that $\delta S^{p+q}x \in F^{2p+2r-1}$. But:

$$\delta S^{p+q}x = S^{p+q}\delta x = \psi^* K_1^* \phi(\delta x \otimes \delta x). \tag{2.55}$$

But since $\delta x \otimes \delta x$ lies in F^{2p+2r}, (2.24b) implies that the right-hand side of (2.55) indeed lies in $F^{2p+2r-1}$, so we are done. □

We remark that in a spectral sequence with products over $\mathbb{Z}/2$, there is an easy survival result for the cup-square. If $\alpha \in E_r$ be given, then α^2 survives to E_{r+1}. Of course this follows from the fact that the differential d_r is a derivation. The result we have just proved is far stronger than this; and, although it refers in its statement only to the product structure of the spectral sequence, it does not seem provable within that context. The only proofs known to this writer involve the cup-1 product K_1, and its behavior with respect to filtration as described by (2.24b).

Theorem 2.17 will be helpful in defining Steenrod squares on the E_∞-term of the spectral sequence. We wish to define homomorphisms:

$$\text{(2.56a)} \qquad Sq^k : E_\infty^{p,q} \to E_\infty^{p,q+k} \qquad \text{if } 0 \le k \le q,$$

$$\text{(2.56b)} \qquad Sq^k : E_\infty^{p,q} \to E_\infty^{p+k-q,2q} \qquad \text{if } q \le k.$$

For the first of these we simply choose $r > \max(p, q+k+1)$ so that both $E_\infty^{p,q} \simeq E_r^{p,q}$ and $E_\infty^{p,q+k} \simeq E_r^{p,q+k}$. Then we define the operation (2.56a) to coincide with the operation (2.46a). As for (2.56b), suppose $k \ge q$ and choose r so that both $r > \max(p, q+1)$ and $2r-1 > \max(p+k-q, 2q+1)$. Then any element in $E_\infty^{p,q}$ can be represented uniquely in the form $[\alpha]_\infty$ with $\alpha \in E_r^{p,q}$. Then by Theorem 2.17, $Sq^k\alpha$ survives to $E_{2r-1}^{p+k-q,2q}$. There it represents an element $[Sq^k\alpha]_\infty \in E_\infty^{p+k-q,2q}$. We set $Sq^k([\alpha]_\infty) = [Sq^k\alpha]_\infty$.

The squaring operations on E_∞ are compatible with those defined on E_0H^*. In fact, with $\pi : E_\infty \to E_0H^*$ the isomorphism of (1.148) we have:

THEOREM 2.22. *The following diagrams commute:*

$$\text{(2.57a)} \qquad \begin{array}{ccc} E_\infty^{p,q} & \xrightarrow{Sq^k} & E_\infty^{p,q+k} \\ \pi\downarrow & & \downarrow\pi \\ E_0^{p,q}H^* & \xrightarrow{E_0(Sq^k)} & E_0^{p,q+k}H^* \end{array} \qquad \text{if } \quad 0 \le k \le q.$$

$$\text{(2.57b)} \qquad \begin{array}{ccc} E_\infty^{p,q} & \xrightarrow{Sq^k} & E_\infty^{p+k-q,2q} \\ \pi\downarrow & & \downarrow\pi \\ E_0^{p,q}H^* & \xrightarrow{E_0(Sq^k)} & E_0^{p+k-q,2q}H^* \end{array} \qquad \text{if } \quad q \le k.$$

PROOF. This is an immediate consequence of (2.28), (1.148), and (2.56). □

Our final task in this section is to study products and Steenrod squares at the E_2 level. In particular, we want to compute the products and squaring operations in terms of the description of E_2 given by (1.149) . For this purpose we will find it useful to divide the squaring operations into two classes, "vertical" and "diagonal", whose definitions in terms of the original operations (2.46) are:

$$\text{(2.58a)} \qquad Sq_V^k = Sq^k : \qquad E_2^{p,q} \to E_2^{p.q+k} \qquad (0 \le k \le q)$$

$$\text{(2.58b)} \qquad Sq_D^k = Sq^{q+k} : \qquad E_2^{p,q} \to E_2^{p+k,2q} \qquad (0 \le k \le p).$$

In terms of the operators $D_*^{*,*}$ of equation (1.150) our result is:

THEOREM 2.23. *Let $x : X_{p,q} \to N$ and $y : X_{r,s} \to N$ be Θ-homomorphisms representing $\alpha \in E_2^{p,q}$ and $\beta \in E_2^{r,s}$, in the sense of (1.149). Then $\alpha \cdot \beta \in E_2^{p+r,q+s}$*

is represented by the composition:

(2.59)

$$X_{p+r,q+s} \xrightarrow{\psi} X_{p+r,q+s} \otimes X_{p+r,q+s} \xrightarrow{D_0^{q,s}(X_{p+r,*};X_{p+r,*})} X_{p+r,q} \otimes X_{p+r,s} \text{———}$$

$$\xrightarrow{D_0^{p,r}(X_{*,q};X_{*,s})} X_{p,q} \otimes X_{r,s} \xrightarrow{x\otimes y} N \otimes N \xrightarrow{\mu} N.$$

If $0 \leq k \leq q$ *then* $Sq_V^k \alpha \in E_2^{p,q+k}$ *is represented by the composition:*

(2.60)

$$X_{p,q+k} \xrightarrow{\psi} X_{p,q+k} \otimes X_{p,q+k} \xrightarrow{D_{q-k}^{q,q}(X_{p,*};X_{p,*})} X_{p,q} \otimes X_{p,q} \xrightarrow{x\otimes x} N \otimes N \xrightarrow{\mu} N.$$

If $0 \leq k \leq p$ *then* $Sq_D^k \alpha \in E_2^{p+k,2q}$ *is represented by the composition:*

(2.61) $$X_{p+k,2q} \xrightarrow{\psi} X_{p+k,2q} \otimes X_{p+k,2q} \xrightarrow{D_0^{q,q}(X_{p+k,*};X_{p+k,*})} X_{p+k,q} \otimes X_{p+k,q} \text{—}$$

$$\xrightarrow{D_{p-k}^{p,p}(X_{*,q};X_{*,q})} X_{p,q} \otimes X_{p,q} \xrightarrow{x\otimes x} N \otimes N \xrightarrow{\mu} N.$$

PROOF. We give the proof for (2.61); the others are similar. With $E_2^{p,q}$ expressed as in (1.149) we suppose $x : X_{p,q} \to N$ represents $\alpha \in E_2^{p,q}$. But the Steenrod squares on E_r were originally defined in terms of the expression (1.144) for E_r. In that language α can be represented by an element w in the double cochain complex $\mathrm{Hom}_\Theta(CX, N)$ of the form $w = x + x'$ where the Θ-homomorphism $x : X_{p,q} \to N$ satisfies $\delta^v x = 0$, and the Θ-homomorphism $x' : X_{p+1,q-1} \to N$ satisfies $\delta^h x = \delta^v x'$. Then $Sq_D^k \alpha = Sq^{k+q}\alpha$ is represented by the cochain $S^{k+q}(w)$. To translate back into the language of (1.149) we have only to pick out from $S^{k+q}(w) \in \oplus_{i \geq p+k} \mathrm{Hom}_\Theta(X_{i,*}, N)$ the component that lies in $\mathrm{Hom}_\Theta(X_{p+k,2q}, N)$. To do this we expand $S^{k+q}(w)$ using (2.14) and ignore any cochain lying in filtration F^{p+k+1}. The expansion can be written

(2.62) $$S^{k+q}(w) = \psi^* K^*_{n-k-q}\phi((x + x') \otimes (x + x')) + \psi^* K^*_{n-k-q+1}\phi(w \otimes \delta w),$$

where $n = p + q$. Since $\delta w \in F^{p+2}$, (2.24b) implies we can ignore the term $\psi^* K^*_{n-k-q+1}\phi(w \otimes \delta w)$. Since $z \in F^{p+1}$, we can also ignore the following terms: $\psi^* K^*_{n-k-q}\phi(x \otimes x')$, $\psi^* K^*_{n-k-q}\phi(x' \otimes x)$, $\psi^* K^*_{n-k-q}\phi(x' \otimes x')$. We conclude that, in the language of (1.149), $Sq^{k+q}\alpha$ is represented by the cochain $\psi^* K^*_{n-k-q}\phi(x \otimes x)$. This cochain is the sum of its components in $\bigoplus_{i \geq p+k} \mathrm{Hom}_\Theta(X_{i,*}, N)$. We use (2.3),(2.4), and (2.6) to show that the component in $\mathrm{Hom}_\Theta(X_{p+k,2q}, N)$ is precisely the composition (2.61). This completes the proof of (2.61). The proofs of (2.59) and (2.60) are similar; in fact, the proof of the latter is written out in [**77**], p.336. In this case one uses the hypothesis that the cup-k product $\{D_k \mid k \geq 0\}$ is "special". □

Theorem 2.23 will be used in Chapter 5 to describe the vertical and diagonal squaring operations on the E_2-term of the change-of-rings spectral sequence, and to derive their properties. Siimilarly, it will be used in Chapter 7 to describe the operations on the E_2-term of the Eilenberg-Moore spectral sequence.

4. Bisimplicial Sets with Group Action

This is a technical section, in which we present a description of the diagonal squaring operations (2.58b) that holds when the bisimplicial coalgebra X is of a special form. We will need this description only for our discussion in Chapter 7 of the diagonal Steenrod squares in the Eilenberg-Moore spectral sequence. We remark that the description of these diagonal squares becomes very much simpler if one is willing to take the left $\mathcal{G}$-space $\mathcal{F}$ of Chapters 6 and 7 to be a point; and the reader who is satisfied with that case can omit this section. The main result is Proposition 2.27.

Throughout this section we fix a simplicial group $\mathcal{G}$, and a simplicial object $\mathcal{W} = \{\mathcal{W}_{p,*} \mid p \geq 0\}$ over the category of right $\mathcal{G}$-spaces. As in Chapter 1, Section 10 we write G for the simplicial vector spaces with basis $\mathcal{G}$, $W_{p,*}$ for the simplicial vector space with basis $\mathcal{W}_{p,*}$, and W for the bisimplicial vector space with basis $\mathcal{W}$. Suppose we are also given a left $\mathcal{G}$-space $\mathcal{F}$. Then we write $\mathcal{W} \times_{\mathcal{G}} \mathcal{F}$ for the bisimplicial set defined by $(\mathcal{W} \times_{\mathcal{G}} \mathcal{F})_{p,*} = \mathcal{W}_{p,*} \times_{\mathcal{G}} \mathcal{F}$. We will write $W \times_G F$ for the bisimplicial vector space with basis $\mathcal{W} \times_{\mathcal{G}} \mathcal{F}$. Then for each $p \geq 0$, $(W \times_G F)_{p,*}$ is the simplicial vector space with basis $\mathcal{W}_{p,*} \times_{\mathcal{G}} \mathcal{F}$. We will also denote this simplicial vector space by $W_{p,*} \times_G F$, this notation being consistent with that introduced after (1.88). Since for each $p \geq 0$, $(W \times_G F)_{p,*}$ is generated freely by a simplicial set, it has the standard structure of a simplicial coalgebra with coproduct $\psi(\sigma) = \sigma \otimes \sigma$ for each $\sigma \in \mathcal{W}_{p,*} \times_{\mathcal{G}} \mathcal{F}$. Clearly, with this coproduct, $W \times_G F$ becomes a bisimplicial coalgebra, and we are interested in applying the theory of this chapter to the spectral sequence of the bisimplicial coalgebra $X = W \times_G F$ with coefficients in $\mathbb{Z}/2$.

The question we ask is the following. Suppose we are given $\alpha \in E_2^{p,q}$ represented in the sense of (1.149) by a cochain $x \in \mathrm{Hom}^q(C(W_{p,*} \times_G F), \mathbb{Z}/2)$. So $\delta_v x = 0$, and there is a $\mathbb{Z}/2$-linear cochain $x' : X_{p+1,q-1} \to N$ satisfying $\delta_h x = \delta_v x'$. In Chapter 6 we will interpret the E_2 term of the Eilenberg-Moore spectral sequence by considering the image

$$\tilde{x} = \phi \bigtriangledown^* (W_{p,*}, F)(x) \tag{2.63}$$

of such cochains x under the composition of cochain mappings

$$\begin{array}{ccc} \mathrm{Hom}(C(W_{p,*} \times_G F), \mathbb{Z}/2) & \xrightarrow{\bigtriangledown^*(W_{p,*}, F)} & \mathrm{Hom}(CW_{p,*} \otimes_{CG} CF, \mathbb{Z}/2) \\ & & \Big\downarrow \phi \\ & & \mathrm{Hom}_{CG}(CW_{p,*}, \overline{\mathrm{Hom}}(CF, \mathbb{Z}/2)), \end{array} \tag{2.64}$$

where the isomorphism ϕ is as in (1.50) and $\bigtriangledown^*(W_{p,*}, F)$ is as in (1.94). We have also $Sq_D^k \alpha$ in $E_2^{p+k,2q}$ represented by a cochain $y \in \mathrm{Hom}^{2q}(C(W_{p+k,*} \times_G F), \mathbb{Z}/2)$, given in terms of x by (2.61). We consider the image:

$$\tilde{y} = \phi \bigtriangledown^* (W_{p+k,*}, F)(y) \tag{2.65}$$

of y under the composition of cochain mappings

(2.66)

$$\begin{array}{ccc} \mathrm{Hom}(C(W_{p+k,*} \times_G F), \mathbb{Z}/2) & \xrightarrow{\nabla^*(W_{p+k,*}, F)} & \mathrm{Hom}(CW_{p+k,*} \otimes_{CG} CF, \mathbb{Z}/2) \\ & & \Big\downarrow \phi \\ & & \mathrm{Hom}_{CG}(CW_{p+k,*}, \overline{\mathrm{Hom}}(CF, \mathbb{Z}/2)). \end{array}$$

The goal of this section is to describe $\tilde{y}$ in terms of $\tilde{x}$. The result appears below as Proposition 2.27.

We begin with a description of $\nabla^*(W_{p+k,*}, F)(y)$.

For each fixed $p \geq 0$ we have been considering the chain complex $CW_{p,*}$ as a module over CG, by the construction (1.82). Therefore the tensor product of chain complexes: $CW_{p,*} \otimes CW_{p,*}$ becomes a module over $CG \otimes CG$. Similarly we write $CW \otimes CW$ for the tensor product of double complexes as described in (2.1), and observe that for each fixed $p \geq 0$, the chain complex $(CW \otimes CW)_{p,*}$ is naturally a module over $CG \otimes CG$.

For each fixed pair of integers $p, s \geq 0$ we restrict $J_s(W, W)$ of (2.4) to the component $CW_{p,*} \otimes CW_{p,*} = (CW \otimes_V CW)_{p,*}$ of $(CW \otimes_V CW)$, obtaining a homomorphism of graded vector spaces:

$$J_s(W, W) : CW_{p,*} \otimes CW_{p,*} \to (CW \otimes CW)_{p+s,*}. \tag{2.67}$$

Similarly for all $p, s \geq 0$ we have:

$$\begin{aligned} &J_s(W \times_G F, W \times_G F) : \\ &\qquad C(W_{p,*} \times_G F) \otimes C(W_{p,*} \times_G F) \to (C(W \times_G F) \otimes C(W \times_G F))_{p+s,*}. \end{aligned} \tag{2.68}$$

LEMMA 2.24. *For each pair of integers $p, s \geq 0$ the homomorphism $J_s(W, W)$ of (2.67) is linear over $CG \otimes CG$.*

PROOF. One needs the facts that:

1. For each fixed $a \geq 0$ and each fixed $g \in G_a$, right multiplication of g upon $W_{*,a}$ is a map of simplicial vector spaces from $W_{*,a}$ to itself;
2. For all $a, b \geq 0$, and for each degeneracy operator $s_\nu = s_{\nu_b} \ldots s_{\nu_1}$, the mapping $W_{*,a} \to W_{*,a+b}$ defined by allowing s_ν to operate vertically is a mapping of simplicial vector spaces;
3. The homomorphisms $D_s(R, S)$ are natural with respect to mappings of the simplicial vector spaces R and S.

The Lemma follows from an easy diagram chase that uses these facts. □

By virtue of Lemma 2.24 we see that for each pair $p, s \geq 0$, the homomorphism $J_s(W, W) : CW_{p,*} \otimes CW_{p,*} \to (CW \otimes CW)_{p+s,*}$, when tensored over $CG \otimes CG$ with the identity on $CF \otimes CF$, passes to a well-defined quotient:

$$(CW_{p,*} \otimes CW_{p,*}) \otimes_{CG \otimes CG} (CF \otimes CF) \to (CW \otimes CW)_{p+s,*} \otimes_{CG \otimes CG} (CF \otimes CF).$$

We note further that as in (1.88), for each integer $p \geq 0$ the chain map

$$\nabla(W_{p,*}, F) : CW_{p,*} \otimes_{CG} CF \to C(W_{p,*} \times_G F)$$

is defined as a quotient of the ordinary shuffle $CW_{p.*} \otimes CF \to C(W_{p,*} \times F)$. In the following diagram we use the symbol ∇_p as an abbreviation for this chain map.

Also we are using the symbol $\nabla_l \otimes \nabla_m$ as an abbreviation for the direct sum of mappings $\oplus_{l+m=p+s}(\nabla_l \otimes \nabla_m)$.

LEMMA 2.25. *For all pairs of integers $p, s \geq 0$ the following diagram of degree-preserving homomorphisms of graded vector spaces commutes:*

$$\begin{array}{ccc}
(CW_{p,*} \otimes_{CG} CF)\otimes(CW_{p,*} \otimes_{CG} CF) & \xrightarrow{\nabla_p\otimes\nabla_p} & C(W_{p,*} \times_G F)\otimes C(W_{p,*} \times_G F) \\
\Big\downarrow{\scriptstyle (1,3,2,4)} & & \\
(CW_{p,*} \otimes CW_{p,*}) \otimes_{CG\otimes CG} (CF \otimes CF) & & \Big\downarrow{\scriptstyle J_s(W \times_G F, W \times_G F)} \\
\Big\downarrow{\scriptstyle J_s(W,W)\otimes(CF\otimes CF)} & & \\
(CW \otimes CW)_{p+s,*} \otimes_{CG\otimes CG} (CF \otimes CF) & & (C(W \times_G F)\otimes C(W \times_G F))_{p+s,*} \\
\Big\downarrow{\scriptstyle (1,3,2,4)} & & \Big\downarrow{\scriptstyle =} \\
\bigoplus_{l+m=p+s} (CW_{l,*}\otimes_{CG}CF)\otimes(CW_{m,*}\otimes_{CG}CF) & \xrightarrow[(\nabla_l\otimes\nabla_m)]{} & \bigoplus_{l+m=p+s} C(W_{l,*}\times_G F)\otimes C(W_{m,*}\times_G F).
\end{array}$$

PROOF. Perhaps the best way to check this is to assume given arbitrary elements $w_a \in W_{p,a}, f_b \in F_b, w_c \in W_{p,c}, f_d \in F_d$. One starts with the specific element $w_a \otimes f_b \otimes w_c \otimes f_d$ in the upper left of the diagram, applies to it both counterclockwise and clockwise compositions, and makes two observations. The first is that for all pairs of "vertical" simplicial operators $s_\nu = s_{\nu_b} \cdots s_{\nu_1}$, $s_{\nu'} = s_{\nu'_d} \cdots s_{\nu'_1}$ the following diagram commutes:

$$\begin{array}{ccc}
W_{p,a} \otimes W_{p,c} & \xrightarrow{s_\nu \otimes s_{\nu'}} & W_{p,a+b} \otimes W_{p,c+d} \\
{\scriptstyle D_s(W_{*,a}W_{*,c})}\Big\downarrow & & \Big\downarrow{\scriptstyle D_s(W_{*,a+b}, W_{*,c+d})} \\
(CW_{*,a} \otimes CW_{*,c})_{p+s} & \xrightarrow[s_\nu \otimes s_{\nu'}]{} & (CW_{*,a+b} \otimes CW_{*,c+d})_{p+s}.
\end{array}$$

This is just a special case of the naturality of the mappings D_s. They are acting horizontally, and so commute with the vertical simplicial operators. The second observation is that, when restricted to the summand $(W \times_G F)_{p,i} \otimes (W \times_G F)_{p,j}$ (here $i = a + b$ and $j = c + d$) the homomorphism $J_s(W \times_G F, W \times_G F)$ is the composition:

$$(W \times_G F)_{p,i} \otimes (W \times_G F)_{p,j} \xrightarrow{(1,3,2,4)} (W_{p,i} \otimes W_{p,j}) \otimes_{\mathcal{G}_i \times \mathcal{G}_j} (F_i \otimes F_j) \xrightarrow{=}$$
$$C_p(W_{*,i} \times W_{*,j}) \otimes_{\mathcal{G}_i \times \mathcal{G}_j} (F_i \otimes F_j) \xrightarrow{D_s(W_{*,i},W_{*,j})\otimes(F_i\otimes F_j)}$$
$$(CW_{*,i}\otimes CW_{*,j})_{p+s}\otimes_{\mathcal{G}_i\times\mathcal{G}_j}(F_i \otimes F_j) \xrightarrow{(1,3,2,4)} (C(W\times_G F)\otimes C(W\times_G F))_{p+s,i+j}.$$

Indeed, this follows from Proposition 1.92 and the naturality of the homomorphisms D_s. The lemma now follows easily from the two observations we have made. $\square$

Now we can carry out the program sketched at the beginning of this section. We assume given cochain $x \in \mathrm{Hom}^q(C(W_{p,*} \times_G F), \mathbb{Z}/2)$ representing element $\alpha \in E_2^{p,q}$ as in (1.149). Associated to x is the cochain $y \in \mathrm{Hom}^{2q}(C(W_{p+k,*} \times_G F), \mathbb{Z}/2)$ representing $Sq_D^k \alpha \in E_2^{p+k,2q}$: y is given in terms of x by (2.61). Then we can describe $\bigtriangledown^*(W_{p+k,*}, F)(y)$ with the aid of the following commutative diagram of homomorphisms of graded vector spaces. In this diagram we are abbreviating the chain map:

$$\bigtriangledown(W_{p+k,*}, F) : CW_{p+k,*} \otimes_{CG} CF \to C(W_{p+k,*} \times_G F)$$

by the symbol $\bigtriangledown_{p+k}$, as we did in the proof of Lemma 2.25. We are abbreviating the chain map:

$$\bigtriangledown_{p+k} \otimes \bigtriangledown_{p+k} : \\ (CW_{p+k,*} \otimes_{CG} CF) \otimes (CW_{p+k,*} \otimes_{CG} CF) \to C((W_{p+k,*} \times_G F) \otimes C(W_{p+k,*} \times_G F))$$

by the symbol $\bigtriangledown^{\otimes}_{p+k}$; and we are abbreviating the chain map

$$\bigtriangledown (W_{p+k,*} \times W_{p+k,*}, F \times F) : \\ C(W_{p+k,*} \times W_{p+k,*}) \otimes_{C(G\times G)} C(F \times F) \to C((W_{p+k,*} \times W_{p+k,*}) \times_{G\times G} (F \times F))$$

by the symbol $\bigtriangledown^{\times}_{p+k}$. Finally, as in Lemma 2.25 we are using the symbol $\bigtriangledown_l \otimes \bigtriangledown_m$ as an abbreviation for the direct sum of mappings $\oplus_{l+m=2p}(\bigtriangledown_l \otimes \bigtriangledown_m)$.

$$\begin{array}{ccc}
CW_{p+k,*} \otimes_{CG} CF & \xrightarrow{\bigtriangledown_{p+k}} & C(W_{p+k,*} \times_G F) \\
\downarrow {\scriptstyle C\psi(W_{p+k,*}) \otimes C\psi(F)} & & {\scriptstyle C(\psi(W_{p+k,*}) \times_G \psi(F))} \downarrow \\
C(W_{p+k,*} \times W_{p+k,*}) \otimes_{C(G\times G)} C(F \times F) & \xrightarrow{\bigtriangledown^{\times}_{p+k}} & C((W_{p+k,*} \times W_{p+k,*}) \times_{G\times G} (F \times F)) \\
\downarrow {\scriptstyle D_0 \otimes D_0} & & {\scriptstyle (1,3,2,4)} \downarrow \\
(CW_{p+k,*} \otimes CW_{p+k,*}) \otimes_{CG\otimes CG} (CF \otimes CF) & & C((W_{p+k,*} \times_G F) \times (W_{p+k,*} \times_G F)) \\
\downarrow {\scriptstyle (1,3,2,4)} & & {\scriptstyle D_0} \downarrow \\
(CW_{p+k,*} \otimes_{CG} CF) \otimes (CW_{p+k,*} \otimes_{CG} CF) & \xrightarrow{\bigtriangledown^{\otimes}_{p+k}} & C((W_{p+k,*} \times_G F) \otimes C(W_{p+k,*} \times_G F)) \\
\downarrow {\scriptstyle (1,3,2,4)} & & {\scriptstyle J_{p-k}(W \times_G F, W \times_G F)} \downarrow \\
(CW_{p+k,*} \otimes CW_{p+k,*}) \otimes_{CG\otimes CG} (CF \otimes CF) & & (C(W \times_G F) \otimes C(W \times_G F))_{2p,*} \\
\downarrow {\scriptstyle J_{p-k}(W, W) \otimes (CF \otimes CF)} & & \Big\| \\
(CW \otimes CW)_{2p,*} \otimes_{CG\otimes CG} (CF \otimes CF) & & = \\
\downarrow {\scriptstyle (1,3,2,4)} & & \Big\downarrow \\
\bigoplus_{l+m=2p} (CW_{l,*} \otimes_{CG} CF) \otimes (CW_{m,*} \otimes_{CG} CF) & \xrightarrow{(\bigtriangledown_l \otimes \bigtriangledown_m)} & \bigoplus_{l+m=2p} C(W_{l,*} \times_G F) \otimes C(W_{m,*} \times_G F).
\end{array}$$

All rectangles in this diagram commute: the top one by (1.89), the middle by (1.93), and the bottom by Lemma 2.25. On the other hand it is clear from (2.61) that $\nabla^*(W_{p+k,*}, F)(y) \in \mathrm{Hom}^{2q}(CW_{p+k,*} \otimes_{CG} CF, \mathbb{Z}/2)$ is the cochain given by the clockwise route around the outside of the diagram, in internal degree $2q$, followed by the composition:

$$\bigoplus_{l+m=2p} (C(W_{l,*} \times_G F) \otimes C(W_{m,*} \times_G F))_{2q} \xrightarrow{\text{proj}} (W_{p,*} \times_G F)_q \otimes (W_{p,*} \times_G F)_q \text{---}$$
$$\xrightarrow{x \otimes x} \mathbb{Z}/2.$$

Consequently, $\nabla^*(W_{p+k,*}, F)(y)$ is also given by the counterclockwise route around the diagram, in internal degree $2q$, followed by the same composition. We summarize:

LEMMA 2.26. *In the spectral sequence of the bisimplicial coalgebra $W \times_G F$, let $\alpha \in E_2^{p,q}$ be represented in the sense of (1.149) by the cochain $x : (W_{p,*} \times_G F)_q \to \mathbb{Z}/2$, and let $Sq_D^k \alpha \in E_2^{p+k,2q}$ be represented by the cochain $y : (W_{p+k,*} \times_G F)_{2q} \to \mathbb{Z}/2$, where y is defined in terms of x by (2.61). Then*

$$\nabla^*(W_{p+k,*}, F)(y) : (CW_{p+k,*} \otimes_{CG} CF)_{2q} \to \mathbb{Z}/2$$

is given by the following composition:

(2.69)

$$(CW_{p+k,*} \otimes_{CG} CF)_{2q} \xrightarrow{C\psi(W_{p+k,*}) \otimes C\psi(F)} (C(W_{p+k,*} \times W_{p+k,*}) \otimes_{C(G \times G)} C(F \times F))_{2q} \text{---}$$
$$\xrightarrow{D_0(W_{p+k,*}) \otimes D_0(F)} ((CW_{p+k,*} \otimes CW_{p+k,*}) \otimes_{CG \otimes CG} (CF \otimes CF))_{2q} \xrightarrow{J_{p-k}(W,W) \otimes (CF \otimes CF)}$$
$$((CW \otimes CW)_{2p,*} \otimes_{CG \otimes CG} (CF \otimes CF))_{2q} \xrightarrow{(1,3,2,4)} \bigoplus_{l+m=2p} ((CW_{l,*} \otimes_{CG} CF) \otimes (CW_{m,*} \otimes_{CG} CF))_{2q}$$
$$\xrightarrow{\oplus_{l,m} \nabla(W_{l,*},F) \otimes \nabla(W_{m,*},F)} \bigoplus_{l+m=2p} (C(W_{l,*} \times_G F) \otimes C(W_{m,*} \times_G F))_{2q} \text{---}$$
$$\xrightarrow{\text{proj}} (W_{p,*} \times_G F)_q \otimes (W_{p,*} \times_G F)_q \xrightarrow{x \otimes x} \mathbb{Z}/2.$$

In terms of the chain map $J_{p-k}^{p,p}(W, W) : CW_{p+k,*} \otimes CW_{p+k,*} \to CW_{p,*} \otimes CW_{p,*}$, as in (1.151), the composition (2.69) can be written more simply in the following form:

(2.70)

$$(CW_{p+k,*} \otimes_{CG} CF)_{2q} \xrightarrow{C\psi(W_{p+k,*}) \otimes C\psi(F)} (C(W_{p+k,*} \times W_{p+k,*}) \otimes_{C(G \times G)} C(F \times F))_{2q} \text{---}$$
$$\xrightarrow{D_0(W_{p+k,*}) \otimes D_0(F)} ((CW_{p+k,*} \otimes CW_{p+k,*}) \otimes_{CG \otimes CG} (CF \otimes CF))_{2q} \xrightarrow{J_{p-k}^{p,p}(W,W) \otimes (CF \otimes CF)}$$
$$((CW_{p,*} \otimes CW_{p,*}) \otimes_{CG \otimes CG} (CF \otimes CF))_{2q} \xrightarrow{(1,3,2,4)} ((CW_{p,*} \otimes_{CG} CF) \otimes (CW_{p,*} \otimes_{CG} CF))_{2q}$$
$$\xrightarrow{\nabla(W_{p,*},F) \otimes \nabla(W_{p,*},F)} (C(W_{p,*} \times_G F) \otimes C(W_{p,*} \times_G F))_{2q} \xrightarrow{\text{proj}} (W_{p,*} \times_G F)_q \otimes (W_{p,*} \times_G F)_q$$
$$\xrightarrow{x \otimes x} \mathbb{Z}/2.$$

The form (2.70) in which we have written $\bigtriangledown^*(W_{p+k,*},F)(y)$ makes it particularly easy to describe the result of applying to this cochain the homomorphism:

$$\phi : \mathrm{Hom}^{2q}(CW_{p+k,*} \otimes_{CG} CF, \mathbb{Z}/2) \to \mathrm{Hom}^{2q}_{CG}(CW_{p+k,*}, \overline{\mathrm{Hom}}(CF,\mathbb{Z}/2)),$$

as in the following proposition.

PROPOSITION 2.27. *In the spectral sequence of the bisimplicial coalgebra*

$$X = W \times_G F,$$

let $\alpha \in E_2^{p,q}$ *be represented as in* (1.149) *by the cochain* $x : (W_{p,*} \times_G F)_q \to \mathbb{Z}/2$, *and let* $Sq^k_D\alpha \in E_2^{p+k,2q}$ *be represented by the cochain* $y : (W_{p+k,*} \times_G F)_{2q} \to \mathbb{Z}/2$, *with* y *defined in terms of* x *by* (2.61). *Let* $\tilde{x} \in \mathrm{Hom}^q_{CG}(CW_{p,*}, \overline{\mathrm{Hom}}(CF,\mathbb{Z}/2))$ *and* $\tilde{y} \in \mathrm{Hom}^{2q}_{CG}(CW_{p+k,*}, \overline{\mathrm{Hom}}(CF,\mathbb{Z}/2))$ *be defined in terms of* x *and* y *by* (2.63) *and* (2.65), *respectively. Then* $\tilde{y}$ *can be expressed in terms of* $\tilde{x}$ *as the composition*

$$(2.71)\quad CW_{p+k,*} \xrightarrow{C\psi(W_{p+k,*})} C(W_{p+k,*} \times W_{p+k,*}) \xrightarrow{D_0(W_{p+k,*},W_{p+k,*})} CW_{p+k,*} \otimes CW_{p+k,*} \text{---}$$
$$\xrightarrow{J^{p,p}_{p-k}(W,W)} CW_{p,*} \otimes CW_{p,*} \xrightarrow{\tilde{x}\otimes\tilde{x}} \overline{\mathrm{Hom}}(CF,\mathbb{Z}/2) \otimes \overline{\mathrm{Hom}}(CF,\mathbb{Z}/2) \xrightarrow{\mu} \overline{\mathrm{Hom}}(CF,\mathbb{Z}/2),$$

where the product μ *on* $\overline{\mathrm{Hom}}(CF,\mathbb{Z}/2)$ *is that defined by* (1.75).

PROOF. We regard each of the chain complexes appearing in (2.71) as a differential right module over CG. Here CG is acting on $C(W_{p+k,*}\times W_{p+k,*})$ through the homomorphism of differential algebras $C\psi_G : CG \to C(G\times G)$, and CG is acting on the next three chain complexes in the sequence (2.71) through the homomorphism of differential algebras (1.81). Then each of the mappings in (2.71) is linear over CG...$C\psi(W_{p+k,*})$ by Proposition 1.98, $D_0(W_{p+k,*}, W_{p+k,*})$ by Proposition 1.100, $J^{p,p}_{p-k}(W,W)$ by Lemma 2.24, and μ by Proposition 1.103. So the composition in (2.71) is indeed linear over CG; we denote it by:

$$\tilde{z} \in \mathrm{Hom}^{2q}_{CG}(CW_{p+k,*}, \overline{\mathrm{Hom}}(CF,\mathbb{Z}/2)).$$

In order to show that $\tilde{z} = \tilde{y}$ we need to show that

$$(2.72)\qquad \zeta(\tilde{z}) = \bigtriangledown^*(W_{p+k,*},F)(y),$$

Here ζ is the inverse of the isomorphism ϕ of (2.66), and $\bigtriangledown^*(W_{p+k,*},F)(y)$ is given by the composition (2.70). We begin our proof of (2.72) by observing that by (1.75), and for any fixed $w \in CW_{p+k,*}$, the homomorphism $\tilde{z}(w) : CF \to \mathbb{Z}/2$ is the composition:

$$(2.73)\qquad CF \xrightarrow{C\psi(F)} C(F\times F) \xrightarrow{D_0(F,F)} (CF\otimes CF) \xrightarrow{\sum_i \tilde{x}(w_i)\otimes\tilde{x}(w_i')} \mathbb{Z}/2.$$

Here we are writing $\sum_i w_i \otimes w_i' \in CW_{p,*} \otimes CW_{p,*}$ for the image of w under the composition of the first three mappings in (2.71). Now, by the definition of ϕ, we have for all $w \in CW_{p+k,*}$ and $f \in CF$ that $(\zeta(\tilde{z}))(w\otimes f) = (\tilde{z}(w)(f)$. So from (2.73) we see that the mapping $\zeta(\tilde{z})$ is the composition:

(2.74)

$$
\begin{aligned}
&(CW_{p+k,*} \otimes_{CG} CF)_{2q} \xrightarrow{C\psi(W_{p+k,*})\otimes C\psi(F)} (C(W_{p+k,*} \times W_{p+k,*}) \otimes_{C(G\times G)} C(F \times F))_{2q} - \\
&\xrightarrow{D_0(W_{p+k,*})\otimes D_0(F)} ((CW_{p+k,*} \otimes CW_{p+k,*}) \otimes_{CG\otimes CG} (CF \otimes CF))_{2q} \xrightarrow{J^{p,p}_{p-k}(W,W)\otimes(CF\otimes CF)} \\
&((CW_{p,*} \otimes CW_{p,*}) \otimes_{CG\otimes CG} (CF \otimes CF))_{2q} \xrightarrow{(1,3,2,4)} ((CW_{p,*} \otimes_{CG} CF) \otimes (CW_{p,*} \otimes_{CG} CF))_{2q} \\
&\xrightarrow{\tilde{x}\otimes CF\otimes\tilde{x}\otimes CF} [(\overline{\mathrm{Hom}}(CF, \mathbb{Z}/2) \otimes_{CG} CF) \otimes (\overline{\mathrm{Hom}}(CF, \mathbb{Z}/2) \otimes_{CG} CF)]_0 \xrightarrow{\mathrm{eval}\otimes\mathrm{eval}} \mathbb{Z}/2 \otimes \mathbb{Z}/2 = \mathbb{Z}/2.
\end{aligned}
$$

Finally it follows from the definition (2.63) that the compositions (2.74) and (2.70) are identical. This completes our proof of (2.72), and the proof of the proposition. $\square$

It is perhaps worth noting that $\delta_v x = 0$, so $\tilde{x}$ is a cocycle in the cochain complex $\mathrm{Hom}_{CG}(CW_{p,*}, \overline{\mathrm{Hom}}(CF, \mathbb{Z}/2))$. So we can think of $\tilde{x}$ as a chain map $\tilde{x} : CW_{p,*} \to \overline{\mathrm{Hom}}(CF, \mathbb{Z}/2)$. Similarly, $\tilde{y} : CW_{p+k,*} \to \overline{\mathrm{Hom}}(CF, \mathbb{Z}/2)$ is a chain map. But $J^{p,p}_{p-k}(W, W)$ is also a chain map, as per our discussion following equation (1.151). So equation (2.71) can be interpreted as expressing $\tilde{y}$ as a composition of chain maps.

5. Application to the Serre Spectral Sequence

In this chapter we have developed a general theory of Steenrod operations in first-quadrant spectral sequences. In the chapters that follow we will apply it to the change-of-rings spectral sequence, and to the Eilenberg-Moore spectral sequence. However, there are other applications that we have not set out in detail. We mention some of these in the closing sections of this chapter.

The first is to the Serre spectral sequence. Andreas Dress has pointed out in [**17**] that the Serre spectral sequence can be obtained as the spectral sequence of a bisimplicial module. We add coproducts to Dress's construction and find that the results of the current chapter can be applied to give Steenrod operations in the spectral sequence. Thus, we recover in our context the original results of Araki [**6**], Kristensen [**34**], and Vasquez-Garcia [**87**].

Dress's construction of the Serre spectral sequence proceeds as follows. To any given Serre fibration $f : \mathcal{E} \to \mathcal{B}$, he associates a bisimplicial set $\mathcal{K}(f)$, whose simplices in bidegree (p, q) are the commutative diagrams of topological spaces and continuous maps of the form:

$$
\begin{array}{ccc}
\Delta_p \times \Delta_q & \xrightarrow{\sigma_{p,q}} & \mathcal{E} \\
\Big\downarrow{\scriptstyle \mathrm{proj}} & & \Big\downarrow{\scriptstyle f} \\
\Delta_p & \xrightarrow[\tau_p]{} & \mathcal{B}.
\end{array}
$$

Here Δ_p and Δ_q are topological p and q-simplices, and "proj" is projection to the first factor. The standard embeddings $\eta_i : \Delta_{p-1} \to \Delta_p$ $(0 \le i \le p)$ and collapses $\lambda_i : \Delta_{p+1} \to \Delta_p$ $(0 \le i \le p)$ induce in the obvious way horizontal faces and degeneracies:

$$d_i^h : \mathcal{K}_{p,q}(f) \to \mathcal{K}_{p-1,q}(f) \ (0 \leq i \leq p),$$
$$s_i^h : \mathcal{K}_{p,q}(f) \to \mathcal{K}_{p+1,q}(f) \ (0 \leq i \leq p).$$

The vertical simplicial operators are defined similarly, by means of the embeddings $\eta_i : \Delta_{q-1} \to \Delta_q \ (0 \leq i \leq q)$ and collapses $\lambda_i : \Delta_{q+1} \to \Delta_q \ (0 \leq i \leq q)$. Dress writes $K(f)$ for the bisimplicial free abelian group with basis $\mathcal{K}(f)$. He shows that the homology of the associated total complex is $H_*CK(f) = H_*(\mathcal{E}, \mathbb{Z})$. He shows further that the E^2-term of the homology spectral sequence of $K(f)$ is:

$$E^2_{p,q} = H_p(\mathcal{B}, H_q(\mathcal{F}_b)).$$

Here $\{H_q(\mathcal{F}_b) \mid b \in \mathcal{B}\}$ is the system of coefficients on $\mathcal{B}$ provided by the q-dimensional integral homology of the fiber over each point.

To adapt the results of Dress to our context, we replace his free abelian groups $K_{p.q}(f)$ by the $\mathbb{Z}/2$ vector spaces with bases $\mathcal{K}_{p,q}(f)$. We write X for the resulting bisimplicial vector space. Dress's work shows that:

$$H^* \operatorname{Hom}(CX, \mathbb{Z}/2) = H^*(\mathcal{E}; \mathbb{Z}/2). \tag{2.75}$$

We look at the cohomology spectral sequence of X as described in Chapter 1, Section 16, with the coefficient module N taken to be $\mathbb{Z}/2$. Equation (2.75) implies that the spectral sequence converges to $H^*(\mathcal{E}; \mathbb{Z}/2)$. Dress's work shows the E_2-term is given by:

$$E_2^{p,q} = H^p(\mathcal{B}, H^q(\mathcal{F}_b; \mathbb{Z}/2)). \tag{2.76}$$

We make X into a bisimplicial coalgebra, by declaring each element of $\mathcal{K}_{p,q}(f)$ "grouplike": $\psi(x) = x \otimes x$ for each $x \in \mathcal{K}_{p,q}(f)$. Then the results of this chapter apply to the spectral sequence. Steenrod operations are defined on each page of the spectral sequence, as described by Theorem 2.15. The operations on E_r are determined by the operations on E_2, as described by Theorem 2.16. The operations commute with differentials, as described by Theorem 2.17. The squaring operations on E_∞ are compatible with those defined on E_0H^*, as described by Theorem 2.22.

Two more steps remain to be taken in order to complete the program. We have not taken them in the present work. The first is to show that the squaring operations on $H^* \operatorname{Hom}(CX, \mathbb{Z}/2)$ defined as in (2.17) agree with those usually defined on $H^*(\mathcal{E}; \mathbb{Z}/2)$, when the identification (2.75) is made. The bisimplicial vector space X admits no obvious augmentation to the simplicial vector space generated by the singular chains on $\mathcal{E}$, so Proposition 2.1 cannot be used here.

The second step is to express the vertical and diagonal squaring operations on the E_2-term of the spectral sequence in the language of the right-hand side of (2.76). Theorem 2.23 should be an effective tool. The vertical operations:

$$Sq_V^k : H^p(\mathcal{B}, H^q(\mathcal{F}_b; \mathbb{Z}/2)) \to H^p(\mathcal{B}, H^{q+k}(\mathcal{F}_b; \mathbb{Z}/2)) \tag{2.77}$$

clearly come from the fact that the coefficient module at each point, $H^*(\mathcal{F}_b; \mathbb{Z}/2)$, is a module over the Steenrod algebra. To describe the diagonal operations:

$$Sq_D^k : H^p(\mathcal{B}, H^q(\mathcal{F}_b; \mathbb{Z}/2)) \to H^{p+k}(\mathcal{B}, H^{2q}(\mathcal{F}_b; \mathbb{Z}/2)) \tag{2.78}$$

one needs a theory of Steenrod squares defined on the cohomology of a space with coefficients a system of commutative algebras over $\mathbb{Z}/2$. One would then apply such a theory to the space $\mathcal{B}$ with coefficient system $H^*(\mathcal{F}_b; \mathbb{Z}/2)$, to obtain an explicit description of the operations (2.78). In case the coefficient system is simple then one has $E_2^{p,q} = H^p(\mathcal{B};\mathbb{Z}/2) \otimes H^q(\mathcal{F};\mathbb{Z}/2)$. In this case one anticipates that the formulas for (2.77) and (2.78) will reduce to those originally obtained in [**6**], [**34**], and [**87**]:

$$Sq_V^k(\alpha \otimes \beta) = \alpha \otimes Sq^k\beta \tag{2.79}$$

$$Sq_D^k(\alpha \otimes \beta) = Sq^k\alpha \otimes \beta^2. \tag{2.80}$$

6. Application to André-Quillen Cohomology

An especially interesting application of the ideas in this chapter has been made by Paul Goerss in his study [**25**] of the André-Quillen cohomology of simplicial augmented algebras.

Given a simplicial object A over the category of commutative, augmented $\mathbb{Z}/2$-algebras, André [**5**] and Quillen [**66**] define the graded cohomology D^*A of A. This cohomology is a representable functor, in the sense that for each $n \geq 0$ there is a simplicial object $K(n)$ over the category of commutative, augmented $\mathbb{Z}/2$-algebras, such that for each object A in the category one has:

$$D^n A = [A, K(n)].$$

Here $[A, B]$ denotes the set of homotopy classes of maps in the category of simplicial algebras. In [**25**] Goerss poses and solves the problem of calculating the André-Quillen cohomology $D^*K(n)$ of these "Eilenberg-Mac Lane" objects. His principal tool is a spectral sequence due to Miller [**49**],[**50**], which for any simplicial algebra A converges to D^*A. Using some ideas from this chapter (as originally set out in [**77**], the reference available at that time), Goerss endows this Miller spectral sequence with products and Steenrod operations. Since these operations commute with differentials, Goerss is able to show that the Miller spectral sequence for $D^*K(n)$ collapses at E_2. In this way he computes $D^*K(n)$ completely. For the very interesting details we refer the reader to [**25**].

CHAPTER 3

Bialgebra Actions on the Cohomology of Algebras

Let Π be a bialgebra, or an algebra with coproducts in the sense of Chapter 1, Section 6. Suppose Ξ an algebra, and M and N a pair of right Ξ-modules. Our object in this chapter is to find sets of conditions on $\Pi, \Xi, M, \text{and} N$ which enable one to define an action of $\overline{\Pi}$, the negative of Π, on the homomorphism group $\mathrm{Hom}_{\Xi}(M, N)$, as well as on the derived functors $\mathrm{Ext}_{\Xi}^{q,*}(M, N)$ for each $q \geq 0$.

In the three sections which follow we will write down three versions of our theory, the distinctions depending on whether Π is a bialgebra, an algebra with coproducts, or a Hopf algebra, and whether the action on $\mathrm{Hom}_{\Xi}(M, N)$ is from the left or the right. We will use the first version to define the "vertical" action of the Steenrod algebra $\mathcal{A}$ on the Eilenberg-Moore spectral sequence; the second to define the vertical action of the bigraded Steenrod algebra $\mathcal{H}$ on the change-of-rings spectral sequence. We will use the third to define the action of the base of an extension of Hopf algebras upon the cohomology of the fiber; an essential step in setting up the change-of-rings spectral sequence. The first and third versions of our theory are special cases of the general version given in Section 2. Nevertheless we have written out the results separately for all three cases, as we think this will make them easier to apply. In each case we have highlighted different aspects of the theory; our choices determined by the applications we are going to make of the separate cases, in the following chapters. We remark finally that a reader interested only in the Eilenberg-Moore spectral sequence need only read the relatively simple version of the theory that is presented in Section 1.

Under appropriate assumptions, the modules $\mathrm{Ext}_{\Xi}^{*,*}(M, N)$ will not only be modules over $\overline{\Pi}$, but also algebras over the bigraded Steenrod algebra $\mathcal{H}$. One of our goals in this chapter is to determine the relationship between the action of $\overline{\Pi}$, on the one hand, and the products and Steenrod squares on the other. Our results will translate in later chapters into a Cartan formula for the vertical Steenrod squares on the Eilenberg-Moore and change-of-rings spectral sequences, and will also give a formula relating vertical and diagonal Steenrod squares in those cases.

1. Left Action by a Bialgebra

In this section we fix a bialgebra Π, a right Π-module M and a left Π-module N, all graded on an abelian group $\mathbb{S}$. We will also assume given a right Π-algebra Ξ; and we will assume that M and N are right Ξ-modules. The main result of this section is Proposition 3.14, which says that if M is a right $\Pi-\Xi$ module in the sense of Definition 1.32, and if the actions of Π and Ξ are "compatible" in the sense of Definition 3.8 below, then formula (3.1) defines an action of $\overline{\Pi}$ upon $\mathrm{Hom}_{\Xi}(M, N)$. The main point will be to show that if $f : M \to N$ is Ξ-linear, then so is $\overline{\pi} \cdot f$. We will then extend our definition of $\overline{\Pi}$-action on $\mathrm{Hom}_{\Xi}(M, N)$ to an action of $\overline{\Pi}$ on $\mathrm{Ext}_{\Xi}^{p,*}(M, N)$ for all $p \geq 0$. This is accomplished in Definition 3.20, with the use of relative homological algebra. Finally, with appropriate extra structures present on M, N, and Ξ, one can define Steenrod operations on $\mathrm{Ext}_{\Xi}^{*,*}(M,N)$ as in (1.126). We will discuss the relationship between these operations, and our newly defined action of $\overline{\Pi}$. The main result in this direction is Proposition 3.21.

We begin by defining a left action of $\overline{\Pi}$ on homomorphism groups of the form $\mathrm{Hom}(M,N)$. Here $\mathrm{Hom}(M,N)$ is the $\mathbb{S}$-graded vector space of homomorphisms linear over the ground field, as in (1.3a), and $\overline{\Pi}$ is the bialgebra negative of Π as defined at the beginning of Chapter 1, Section 4. Our definition generalizes (1.24).

If π is an element of Π, we will write $\overline{\pi}$ for the corresponding element of $\overline{\Pi}$.

DEFINITION 3.1. For f in $\mathrm{Hom}(M,N)$ and $\pi \in \Pi$ we define $\overline{\pi} \cdot f \in \mathrm{Hom}(M,N)$ by writing

$$(\overline{\pi} \cdot f)(m) = \sum_i (-1)^{|f(m)||\pi_i'|} \pi_i f(m\pi_i') \tag{3.1}$$

for all $m \in M$. Here we have chosen an expansion $\psi(\pi) = \sum_i \pi_i \otimes \pi_i'$ for the coproduct.

Our definition of $\overline{\pi} \cdot f$ is clearly independent of the choice of expansion for $\psi(\pi)$.

REMARK 3.2. Equation (3.1) employs the kind of shorthand that has been described in Remark 1.7. Each component of $\overline{\pi} \cdot f$ is being defined in terms of the components of f.

The signs in (3.1), like the signs in the formulae which follow, are chosen by the principle that whenever two symbols are interchanged, one inserts a factor of -1 raised to a power that is the product of the degrees of the symbols. The signs are necessary for the following propositions.

PROPOSITION 3.3. *The definition* (3.1) *gives* $\mathrm{Hom}(M,N)$ *the structure of a left* $\overline{\Pi}$*-module.*

In fact, the reader will check that the mapping $\overline{\Pi} \otimes \mathrm{Hom}(M,N) \to \mathrm{Hom}(M,N)$ defined by (3.1) preserves degree, as required by our definition of a module over an algebra. The verification that $(\sigma\tau)\cdot f = \sigma\cdot(\tau\cdot f)$ for all $\sigma, \tau \in \overline{\Pi}$ is a straightforward computation that uses the fact that the product $\mu : \Pi \otimes \Pi \to \Pi$ is a morphism of coalgebras.

The action of $\overline{\Pi}$ that we have just constructed is natural, in the following sense:

PROPOSITION 3.4. *Let* Π *be a bialgebra, and* $g : M' \to M$ *a homomorphism of right* Π*-modules homogeneous of any degree. Let* $h : N \to N'$ *be a homomorphism of left* Π*-modules homogeneous of any degree. Then the induced map*

$$\langle g, h\rangle : \mathrm{Hom}(M,N) \to \mathrm{Hom}(M',N')$$

is a homomorphism of left $\overline{\Pi}$*-modules.*

REMARK 3.5. The proof of this proposition is completely straightforward, but we note that the naturality of the Π-action with respect to the homomorphisms $h : N \to N'$ depends on the sign convention (1.22).

We will need to know how the $\overline{\Pi}$ action we have just constructed relates to the pairing ν of (1.7a). So we suppose given a pair of bialgebras Π, Π'; a right Π-module M, and a right Π'-module M'; a left Π-module N and a left Π'-module N'. Then from Proposition 3.3 we have actions of $\overline{\Pi}$ on $\mathrm{Hom}(M,N)$, of $\overline{\Pi}'$ on $\mathrm{Hom}(M',N')$, and of $\overline{\Pi} \otimes \overline{\Pi}'$ on $\mathrm{Hom}(M \otimes M', N \otimes N')$.

PROPOSITION 3.6. *Under the hypotheses above, the pairing of* (1.7a)

$$\nu : \mathrm{Hom}(M, N) \otimes \mathrm{Hom}(M', N',) \to \mathrm{Hom}(M \otimes M', N \otimes N')$$

is a homomorphism of $\overline{\Pi} \otimes \overline{\Pi}'$*- modules.*

This is an easy consequence of the definitions; mainly one must keep track of the signs.

Now we bring in a second algebra Ξ, assumed a right algebra over the bialgebra Π, as in Definition 1.26. Suppose both M and N of Definition 3.1 have right Ξ actions, in addition to their actions by Π. We ask, what relations must we assume between the Π and Ξ actions on M and N in order that $\mathrm{Hom}_\Xi(M, N)$ be a $\overline{\Pi}$-submodule of $\mathrm{Hom}(M, N)$?

We remind the reader of the notion of a $\Pi - \Xi$ module, as in Definition 1.32.

EXAMPLE 3.7. Suppose M a right $\Pi - \Xi$ module on which Ξ acts from the right, and R a right $\Pi - \Xi$ module on which Ξ acts from the left. As in Proposition 1.33 we observe that the diagonal action of Π on $M \otimes R$ passes to an action of Π on the quotient $M \otimes_\Xi R$. As in (1.24) we get a dual, left action of $\overline{\Pi}$ on the vector space $\mathrm{Hom}(M \otimes_\Xi R, l)$. On the other hand we can set $N = \overline{\mathrm{Hom}}(R, l)$ and observe that N becomes a right Ξ-module (as in (1.25)), and a left Π-module (by a similar construction). Now it is easy to see that the adjunction isomorphism $\phi : \mathrm{Hom}(M \otimes R, l) \to \mathrm{Hom}(M, \overline{\mathrm{Hom}}(R, l))$ of (1.4) restricts to an isomorphism of graded vector spaces:

$$\mathrm{Hom}(M \otimes_\Xi R, l) \xrightarrow{\phi} \mathrm{Hom}_\Xi(M, \overline{\mathrm{Hom}}(R, l)) \cong \mathrm{Hom}_\Xi(M, N). \tag{3.2}$$

But we have just observed that the left hand side of (3.2) is a left $\overline{\Pi}$-module, so $\mathrm{Hom}_\Xi(M, N)$ acquires from (3.2) a left $\overline{\Pi}$ action. It is easy to check that this action is the restriction of the action of $\overline{\Pi}$ upon $\mathrm{Hom}(M, N)$ that we defined in (3.1). So in this special case we have succeeded in our goal of defining $\mathrm{Hom}_\Xi(M, N)$ as a $\overline{\Pi}$-submodule of $\mathrm{Hom}(M, N)$.

We will want to deal with Ξ-modules N that do not have a natural description as the duals of other modules. So we extract from the example just given the property of N that makes it work.

DEFINITION 3.8. Let Π be a bialgebra, and Ξ a right Π-algebra. Suppose N a left Π-module with action $\Pi \otimes N \xrightarrow{\sigma_N} N$, that is also a right Ξ-module with action $N \otimes \Xi \xrightarrow{\sigma'_N} N$. We say the actions of Π and Ξ are compatible if the following diagram commutes:

$$
\begin{array}{ccccc}
\Pi\otimes N\otimes\Xi & \xrightarrow{\ \sigma_N\otimes\Xi\ } & N\otimes\Xi & \xrightarrow{\ \sigma'_N\ } & N \\
{\scriptstyle \psi\otimes N\otimes\Xi}\Big\downarrow & & & & \Big\uparrow \\
\Pi\otimes\Pi\otimes N\otimes\Xi & & & & \sigma_N \\
{\scriptstyle (1,3,4,2)}\Big\downarrow & & & & \Big| \\
\Pi\otimes N\otimes\Xi\otimes\Pi & \xrightarrow[\ \Pi\otimes N\otimes\sigma_\Xi\ ]{} & \Pi\otimes N\otimes\Xi & \xrightarrow[\ \Pi\otimes\sigma'_N\ ]{} & \Pi\otimes N.
\end{array}
\tag{3.3}
$$

We will speak of a module N satisfying this definition as "a module with compatible actions of Π on the left and Ξ on the right". If N and N' are two such modules, then by a homomorphism $f : N \to N'$ we mean a degree-preserving map that is both a homomorphism of Π-modules and a homomorphism of Ξ-modules.

We will find it useful to express the commutativity of (3.3) in terms of elements:

$$
(\pi n)\xi = \sum_i (-1)^{|\pi'_i|(|n|+|\xi|)}\pi_i(n(\xi\cdot\pi'_i))
\tag{3.4}
$$

for all $n \in N, \xi \in \Xi, \pi \in \Pi$. Here the expression $\xi \cdot \pi'_i$ refers to the right action of Π upon Ξ.

EXAMPLE 3.9. Suppose Π a bialgebra and Ξ a right Π-algebra. Suppose R a right $\Pi - \Xi$ module with left action of Ξ. Set $N = \overline{\mathrm{Hom}}(R, l)$ with dual actions of Π and Ξ, as in Example 3.7. We claim that the actions of Π and Ξ on N are compatible. One verifies (3.4) by evaluating both sides on an arbitrary $r \in R$, paying attention to the sign in (1.24). One uses the hypothesis that R is a right $\Pi - \Xi$ module, and the cocommutativity of Π. Our global assumption that actions of algebras on modules are degree-preserving is needed to make the signs work out correctly.

The next few propositions develop some elementary properties of modules with compatible actions of Π and Ξ. They are analogous to Propositions 1.34 through 1.37, in which we developed similar properties of $\Theta - \Xi$ modules.

PROPOSITION 3.10. *Let Δ be a bialgebra, Ξ a right Δ-algebra, and N a module with compatible actions of Δ on the left and Ξ on the right. Suppose given a bialgebra homomorphism $\beta : \Pi \to \Delta$. Then with Π acting both on the right of Ξ and the left of N through β, Ξ becomes a Π-algebra, and the actions of Π and Ξ on N are compatible.*

The first part is already in Proposition 1.27 and the second part is easy.

Proposition 3.10 gives one way to "pull back" a module with compatible actions. The following proposition gives another.

PROPOSITION 3.11. *Let Π be a bialgebra, Υ a right Π-algebra. Let N be a left Π-module and a right Υ-module in such a way that the actions of Π and Υ are compatible. Suppose $\beta : \Xi \to \Upsilon$ a homomorphism of right Π-algebras. If we allow*

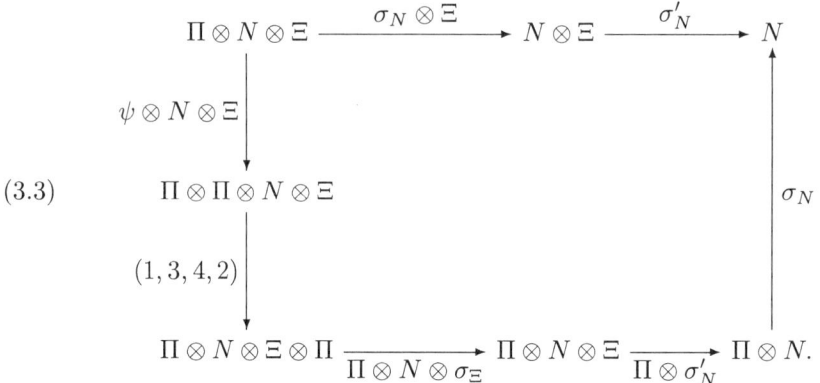

Ξ to act on the right of N through β, then the actions of Π and Ξ upon N are compatible.

We have further:

PROPOSITION 3.12. *Suppose given:*

1. *A bialgebra Π, a right Π-algebra Ξ, and a module N with compatible actions of Π on the left and Ξ on the right;*
2. *A bialgebra Π', a right Π'-algebra Ξ', and a module N' with compatible actions of Π' on the left and Ξ' on the right.*

Then $\Xi \otimes \Xi'$ is a right algebra over the bialgebra $\Pi \otimes \Pi'$, and $N \otimes N'$ is a module with compatible actions of $\Pi \otimes \Pi'$ on the left and $\Xi \otimes \Xi'$ on the right.

The first part of the conclusion is already in Proposition 1.28; the second is an easy diagram chase.

As a corollary (and by analogy with Proposition 1.37) we have:

PROPOSITION 3.13. *Let Π be a bialgebra and Ξ a right bialgebra over Π. Suppose N and N' a pair of modules each with compatible actions of Π on the left and Ξ on the right. Then the diagonal actions of Π and Ξ on $N \otimes N'$ are compatible.*

PROOF. We have from Proposition 3.12 that the actions of $\Pi \otimes \Pi$ and $\Xi \otimes \Xi$ on $N \otimes N'$ are compatible. Then we allow Π to act diagonally on both $\Xi \otimes \Xi$ and $N \otimes N'$, and use Proposition 3.10 to conclude that the actions of Π and $\Xi \otimes \Xi$ on $N \otimes N'$ are compatible. Finally, since the coproduct $\psi_\Xi : \Xi \to \Xi \otimes \Xi$ is a homomorphism of Π-algebras, we can appeal to Proposition 3.11 to complete the proof. □

We have introduced the idea of modules with compatible actions of Π and Ξ so that we can state and prove the following result.

PROPOSITION 3.14. *Let Π be a bialgebra, and Ξ a right Π-algebra. Let M and N be a pair of right Ξ-modules. Suppose given actions of Π on the right of M, and on the left of N, in such a way that:*

1. *M is a right $\Pi - \Xi$ module in the sense of Definition 1.32;*
2. *the actions of Π and Ξ on N are compatible in the sense of Definition 3.8.*

Let $\mathrm{Hom}(M, N)$ *is regarded as a left $\overline{\Pi}$-module, as in* (3.1)*. Then* $\mathrm{Hom}_\Xi(M, N)$ *is a sub-$\overline{\Pi}$ module. Thus,* (3.1) *defines an action:*

$$\overline{\Pi} \otimes \mathrm{Hom}_\Xi(M, N) \to \mathrm{Hom}_\Xi(M, N). \tag{3.5}$$

This action is natural in the sense that if $g : M' \to M$ is a homomorphism of $\Pi - \Xi$ modules, and if $h : N \to N'$ is a homomorphism of modules with compatible actins of Π on the left and Ξ on the right, then the induced mapping:

$$\langle g, h\rangle : \mathrm{Hom}_\Xi(M, N) \to \mathrm{Hom}_\Xi(M', N')$$

is a homomorphism of left $\overline{\Pi}$-modules.

PROOF. We must check that

$$(\overline{\pi} \cdot f)(m\xi) = [(\overline{\pi} \cdot f)m]\xi \tag{3.6}$$

for all $f \in \mathrm{Hom}_\Xi(M, N)\,, \pi \in \Pi\,, m \in M\,, \xi \in \Xi$. Given $\pi \in \Pi$, choose an expansion $\sum_i \pi_i \otimes \pi_i' \otimes \pi_i''$ for its image under the mapping $(\psi \otimes \Pi)\psi : \Pi \to \Pi \otimes \Pi \otimes \Pi$. Using

(3.1) and our hypothesis that M is a right $\Pi - \Xi$ module we see that the left-hand side of (3.6) is:

$$(\overline{\pi} \cdot f)(m\xi) = \sum_i (-1)^{|f(m\xi)||\pi_i''|+|f(m)||\pi_i'|}\pi_i\big(f(m\pi_i')(\xi \cdot \pi_i'')\big).$$

Here the expression $\xi \cdot \pi_i''$ refers to the right action of Π on Ξ. But one uses (3.1), (3.4), and the implicitly assumed cocommutativity of Π to show that the right-hand side of (3.6) is also given by this sum. Our global assumption that actions of algebras on modules are degree-preserving is needed to make the signs work out correctly. Naturality of the action (3.5) in the variables M, N is implied by Proposition 3.4. □

EXAMPLE 3.15. Suppose Π a bialgebra and Ξ a right Π-algebra. Suppose R a right $\Pi - \Xi$ module with left action of Ξ. Set $N = \overline{\mathrm{Hom}}(R, l)$ with dual actions of Π and Ξ. According to Example 3.9, the actions of Π and Ξ on N are compatible. If now we are also given a right $\Pi - \Xi$ module M with right Ξ-action, we have satisfied the hypotheses of Proposition 3.14. So we obtain from that proposition a left action of $\overline{\Pi}$ upon $\mathrm{Hom}_\Xi(M, N)$. It is easy to check that this action is the same as that resulting from (3.2).

The action we have constructed in Proposition 3.14 is natural with respect to mappings of the algebra Ξ in the following sense. Let Π be a bialgebra, and Υ a right Π-algebra. Let M and N be a pair of right Υ-modules. Suppose given actions of Π on the right of M, and on the left of N, in such a way that:

1. M is a right $\Pi - \Upsilon$ module;
2. the actions of Π and Υ on N are compatible in the sense of Definition 3.8.

Suppose finally we are given a homomorphism $\beta : \Xi \to \Upsilon$ of Π-algebras. Then M and N become right Ξ-modules through the action of β, and by Propositions 1.35 and 3.11 they and satisfy conditions 1 and 2 above with Υ replaced by Ξ. Thus, Proposition 3.14 gives us actions of $\overline{\Pi}$ on both $\mathrm{Hom}_\Upsilon(M, N)$ and $\mathrm{Hom}_\Xi(M, N)$.

PROPOSITION 3.16. *Under these assumptions the induced* $\beta^* : \mathrm{Hom}_\Upsilon(M, N) \to \mathrm{Hom}_\Xi(M, N)$ *is a homomorphism of* $\overline{\Pi}$*-modules.*

The proof is immediate from the definitions.

We study the relationship of the action of $\overline{\Pi}$ that we have constructed in Proposition 3.14 with the pairing ν of (1.23). So we suppose we have two bialgebras Π and Π'; we suppose Ξ a right Π-algebra and Ξ' a right Π'-algebra. Let M, N be a pair of right modules over Ξ, and M', N' a pair of right modules over Ξ'. We assume given actions of Π on the right of M, of Π' on the right of M'; and actions of Π on the left of N, of Π' on the left of N', in such a way that:

1. M is a right $\Pi - \Xi$ module and M' is a right $\Pi' - \Xi'$ module;
2. the actions of Π and Ξ on N are compatible; and the actions of Π' and Ξ' on N' are compatible.

Then from Propostions 1.36 and 3.12 we have that $M \otimes M'$ is a right $\Pi \otimes \Pi' - \Xi \otimes \Xi'$ module, and that the actions of $\Pi \otimes \Pi'$ and $\Xi \otimes \Xi'$ on $N \otimes N'$ are compatible. Thus, Proposition 3.14 gives us not only actions of $\overline{\Pi}$ on $\mathrm{Hom}_\Xi(M, N)$ and of $\overline{\Pi}'$ on $\mathrm{Hom}_{\Xi'}(M', N')$, but also an action of $\overline{\Pi} \otimes \overline{\Pi}'$ on $\mathrm{Hom}_{\Xi \otimes \Xi'}(M \otimes M', N \otimes N')$.

PROPOSITION 3.17. *Under the hypotheses above, the pairing of* (1.23a)

$$\nu : \mathrm{Hom}_{\Xi}(M, N) \otimes \mathrm{Hom}_{\Xi'}(M', N',) \to \mathrm{Hom}_{\Xi\otimes\Xi'}(M \otimes M', N \otimes N') \tag{3.7}$$

is a homomorphism of $\overline{\Pi} \otimes \overline{\Pi}'$*-modules.*

This follows at once from Proposition 3.6.

Now we can study the relation of the $\overline{\Pi}$-action constructed in Proposition 3.14 to the pairing ρ of (1.30). We suppose that Π is a bialgebra, and that Ξ is a right Π-bialgebra. Let M, N be a pair of right modules over Ξ, having actions of Π satisfying the hypothesis of Proposition 3.14. Let M', N' be another such pair of right modules over Ξ. We then give $M \otimes M'$ and $N \otimes N'$ the diagonal actions of both Ξ and Π. Then we have from from Propositions 1.37 and 3.13 that $M \otimes M'$ is a Π-Ξ module, and that the actions of Π and Ξ on $N \otimes N'$ are compatible. Thus Proposition 3.14 gives $\overline{\Pi}$ actions not only on $\mathrm{Hom}_{\Xi}(M, N)$ and $\mathrm{Hom}_{\Xi}(M', N')$, but also on $\mathrm{Hom}_{\Xi}(M \otimes M', N \otimes N')$. We wish to study the homomorphism ρ of (1.30), which in the present context is defined by the following diagram:

$$\begin{array}{ccc} \mathrm{Hom}_{\Xi}(M, N) \otimes \mathrm{Hom}_{\Xi}(M', N') & & \\ \downarrow \nu & \searrow \rho & \\ \mathrm{Hom}_{\Xi\otimes\Xi}(M \otimes M', N \otimes N') & \xrightarrow[\psi^*]{} & \mathrm{Hom}_{\Xi}(M \otimes M', N \otimes N'), \end{array} \tag{3.8}$$

where $\psi : \Xi \to \Xi \otimes \Xi$ is the coproduct. Our aim is to show that ρ is a homomorphism of $\overline{\Pi}$-modules, when $\overline{\Pi}$ is acting diagonally on $\mathrm{Hom}_{\Xi}(M, N) \otimes \mathrm{Hom}_{\Xi}(M', N')$.

We begin by remarking that Proposition 3.17 gives actions of $\overline{\Pi} \otimes \overline{\Pi}$ on both domain and codomain of ν, and says that ν is linear over $\overline{\Pi} \otimes \overline{\Pi}$. So we have immediately:

LEMMA 3.18. *The mapping ν of* (3.8) *is a homomorphism of $\overline{\Pi}$-modules, if $\overline{\Pi}$ acts through its coproduct on domain and codomain.*

But there is another way to define an action of $\overline{\Pi}$ on $\mathrm{Hom}_{\Xi\otimes\Xi}(M \otimes M', N \otimes N')$. By Proposition 1.29 we can regard $\Xi \otimes \Xi$ as a right Π-algebra under diagonal action of Π. Then using Propositions 1.36 and 1.34 we can regard $M \otimes M'$ as a right $\Pi - (\Xi \otimes \Xi)$ module. Similarly, using Propositiions 3.12 and 3.10 we can regard $N \otimes N'$ as a module with compatible actions of Π and $\Xi \otimes \Xi$, where Π is acting diagonally. Proposition 3.14 now gives a $\overline{\Pi}$ action on $\mathrm{Hom}_{\Xi\otimes\Xi}(M \otimes M', N \otimes N')$. It is easy to check that this action is the same as that referred to in Lemma 3.18. Now it follows from Proposition 3.16 that ψ^*_{Ξ} of (3.8) is a homomorphism of $\overline{\Pi}$-modules. Here, of course, we are using our hypothesis that the coproduct $\psi : \Xi \to \Xi \otimes \Xi$ is a homomorphism of Π-algebras. So we have proved:

PROPOSITION 3.19. *Suppose Π a bialgebra, and Ξ a right Π-bialgebra. Let M, N be a pair of right modules over Ξ, having actions of Π satisfying the hypothesis of Proposition 3.14. Let M', N' be another such pair of right modules over Ξ. Then the pairing ρ of* (3.8) *is a homomorphism of $\overline{\Pi}$-modules if we understand $\overline{\Pi}$ to be acting diagonally on the domain of ρ.*

We return now to the general situation envisioned in Proposition 3.14. We wish to extend the action of $\overline{\Pi}$ on $\mathrm{Hom}_{\Xi}(M, N)$ that we constructed there to an action of $\overline{\Pi}$ upon $\mathrm{Ext}_{\Xi}^{p,*}(M,N)$ for each $p \geq 0$. We recall from Chapter 1, Definition 1.41, the notion of a relatively projective $\Pi - \Xi$ module; also from Definition 1.45 the notion of a resolution of an arbitrary $\Pi - \Xi$ module M by relative projectives. This definition says in particular that a resolution of a $\Pi - \Xi$ module M by relative projectives is also a projective resolution of M as a Ξ-module, in the ordinary sense. So for any right Ξ-module N we have:

$$\mathrm{Ext}_{\Xi}^{p,*}(M,N) = H^{p,*}\,\mathrm{Hom}_{\Xi}(\mathcal{P}(M),N) \tag{3.9}$$

an isomorphism of $\mathbb{S}$-graded vector spaces for each $p \geq 0$. But suppose N also has a left Π-action that is compatible with the action of Ξ, in the sense of Definition 3.8. Then by Proposition 3.14, $\mathrm{Hom}_{\Xi}(\mathcal{P}(M),N)$ becomes a cochain complex over the category of left $\overline{\Pi}$-modules. So (3.9) defines a left $\overline{\Pi}$-action on $\mathrm{Ext}_{\Xi}^{p,*}(M,N)$, for each $p \geq 0$. By the uniqueness of the projective resolution up to chain homotopy equivalence of $\Pi - \Xi$ complexes (Remark 1.46), the action of $\overline{\Pi}$ is well defined. We summarize:

DEFINITION 3.20. Let Π be a bialgebra, Ξ a right Π-algebra, M a right $\Pi - \Xi$ module, and N a module with compatible actions of Π on the left and Ξ on the right. Let $\mathcal{P}(M) \to M \to 0$ be any resolution of M by relative projectives in the category of $\Pi - \Xi$ modules. Then (3.9) defines a left action of $\overline{\Pi}$ on $\mathrm{Ext}_{\Xi}^{p,*}(M,N)$, for each $p \geq 0$, that is independent of the choice of resolution.

In the remainder of this section we fix the ground field to be $l = \mathbb{Z}/2$ and we suppose $\Pi, \Xi, M,$ and$N$ to be as in Definition 3.20. We ask: suppose Ξ, M and N have the additional structures necessary for the definition of products and Steenrod squares on $\mathrm{Ext}_{\Xi}^{*,*}(M,N)$, as in Proposition 1.120. How do these operations relate to the action of $\overline{\Pi}$ that we have just constructed? To define the products and Steenrod squares we will have to endow Ξ and M with coproducts, and N with a product, and we will have to assume the new structures on M and N are linear over Ξ. To relate the action of $\overline{\Pi}$ to the products and Steenrod squares, we will have to assume that the coproducts on Ξ and M, and the product on N, are linear over Π. In this connection we remind the reader of Definitions 1.26.

In the following proposition we will regard $\mathrm{Ext}_{\Xi}^{*,*}(M,N)$ as an algebra graded on $\mathbb{Z} \times \mathbb{S}$, where the first factor gives homological degree. We will also regard $\overline{\Pi}$ as a bialgebra graded on $\mathbb{Z} \times \mathbb{S}$: it is concentrated in the degrees $0 \times \mathbb{S}$. With these conventions, we can regard equation (3.9) as defining $\mathrm{Ext}_{\Xi}^{*,*}(M,N)$ as a left $\overline{\Pi}$-module.

PROPOSITION 3.21. *Let Π be a bialgebra, Ξ a right Π-Hopf algebra. Suppose M a right $\Pi - \Xi$ module, and N a module with compatible actions of Π on the left and Ξ on the right. Suppose in addition that M is a coalgebra and N a commutative algebra, in such a way that:*

1. *M is a Ξ-coalgebra, and a Π-coalgebra;*
2. *N is a Ξ-algebra and a Π-algebra.*

With the action of $\overline{\Pi}$ on $\mathrm{Ext}_{\Xi}^{*,*}(M,N)$ defined by (3.9), and with the product on $\mathrm{Ext}_{\Xi}^{*,*}(M,N)$ as in (1.125), then $\mathrm{Ext}_{\Xi}^{*,*}(M,N)$ becomes a commutative $\overline{\Pi}$-algebra. Further, with the Steenrod operations on $\mathrm{Ext}_{\Xi}^{*,*}(M,N)$ defined as in (1.126) we have

$$\overline{\pi} Sq^k \alpha = Sq^k[V(\overline{\pi})\alpha] \tag{3.10}$$

for all $\overline{\pi} \in \overline{\Pi}$, *all* $\alpha \in \mathrm{Ext}_{\Xi}^{*,*}(M, N)$ *and all* $k \geq 0$. *Here* V *is the Verschiebung of* (1.17).

PROOF. Let $\mathcal{P}(M) \to M \to 0$ be a resolution of the right $\Pi - \Xi$ module M by relative projectives. Then by Proposition 1.37 the tensor product of this augmented chain complex with itself, $\mathcal{P}(M) \otimes \mathcal{P}(M) \to M \otimes M \to 0$, is also an augmented chain complex over the category of right $\Pi - \Xi$ modules. It is here, in our application of Proposition 1.37, that we need our hypothesis that the coproduct on Ξ is linear over Π. But $\mathcal{P}(M) \to M \to 0$ has a contracting homotopy that is linear over Π in each homological degree, so it is easy to see that $\mathcal{P}(M) \otimes \mathcal{P}(M) \to M \otimes M \to 0$ also has such a contracting homotopy. From this fact, and from our assumptions that the coproduct $\psi : M \to M \otimes M$ is linear over both Ξ and Π, it follows that one can construct a "homological cup-k product", in the sense of Definition 1.121: $\{\mathcal{D}_k : \mathcal{P}(M) \to \mathcal{P}(M) \otimes \mathcal{P}(M) \mid k \geq 0\}$ in which each mapping $\mathcal{D}_k$, when restricted to any particular homological degree, is a homomorphism of $\Pi - \Xi$ modules. The dual mappings, as in (1.128), are for all $k, n \geq 0$:

$$\mathcal{D}_k^* : \mathrm{Hom}_{\Xi}^{n+k}(\mathcal{P}(M) \otimes \mathcal{P}(M), N) \to \mathrm{Hom}_{\Xi}^{n}(\mathcal{P}(M), N). \tag{3.11}$$

Now Proposition 3.14 gives left $\overline{\Pi}$ actions on both sides of (3.11), and asserts that D_k^* is linear over $\overline{\Pi}$. To complete our description of the product and Steenrod operations on $\mathrm{Ext}_{\Xi}^{*,*}(M, N)$ we review the mapping of cochain complexes

$$\phi : \mathrm{Hom}_{\Xi}(\mathcal{P}(M), N) \otimes \mathrm{Hom}_{\Xi}(\mathcal{P}(M), N) \to \mathrm{Hom}_{\Xi}(\mathcal{P}(M) \otimes \mathcal{P}(M), N) \tag{3.12}$$

defined as in (1.130); only now it is important for us to add the observation that ϕ can be written as the composition:

$$\begin{array}{ccc} \mathrm{Hom}_{\Xi}(\mathcal{P}(M), N) \otimes \mathrm{Hom}_{\Xi}(\mathcal{P}(M), N) & \xrightarrow{\rho} & \mathrm{Hom}_{\Xi}(\mathcal{P}(M) \otimes \mathcal{P}(M), N \otimes N). \\ & \searrow^{\phi} & \downarrow \mu_* \\ & & \mathrm{Hom}_{\Xi}(\mathcal{P}(M) \otimes \mathcal{P}(M), N). \end{array} \tag{3.13}$$

Here ρ is as in Proposition 3.19, and $\mu : N \otimes N \to N$ is the Ξ-linear multiplication on N. From Proposition 3.19 and our assumption that μ is linear over Π it follows that ϕ, when restricted to any particular homological degree, is a homomorphism of $\overline{\Pi}$-modules. Now the cochain operation

$$m : \mathrm{Hom}_{\Xi}(\mathcal{P}(M), N) \otimes \mathrm{Hom}_{\Xi}(\mathcal{P}(M), N) \to \mathrm{Hom}_{\Xi}(\mathcal{P}(M), N), \tag{3.14}$$

defined as in (1.131) by the formula $m = \mathcal{D}_0^* \phi$, is in each homological degree a composition of $\overline{\Pi}$-linear maps, and so is $\overline{\Pi}$-linear. It follows that $\mathrm{Ext}_{\Xi}^{*,*}(M, N)$ is a $\overline{\Pi}$-algebra: this is the first part of our theorem. To prove (3.10), let $\alpha \in \mathrm{Ext}_{\Xi}^{p,*}(M, N)$ be represented by a Ξ-linear cocycle $x : \mathcal{P}_p(M) \to N$. Let $\pi \in \Pi$ be given, with coproduct (1.17). Bearing in mind the $\overline{\Pi}$-linearity of the mappings (3.11) and (3.12) we have from the definition (1.132) that for each $\pi \in \Pi$, $\overline{\pi} Sq^k \alpha$

is represented by the cocycle

$$\begin{aligned}\overline{\pi}\mathcal{D}^*_{p-k}\phi(x\otimes x) &= \mathcal{D}^*_{p-k}\phi(\overline{\pi}(x\otimes x))\\ &= \sum_i \mathcal{D}^*_{p-k}\phi(\overline{\pi}_i x\otimes\overline{\pi}'_i x+\overline{\pi}'_i x\otimes\overline{\pi}_i x)+\mathcal{D}^*_{p-k}\phi(V(\overline{\pi})x\otimes V(\overline{\pi})x)\\ &= \sum_i (\mathcal{D}^*_{p-k}+\mathcal{D}^*_{p-k}T^*)\phi(\overline{\pi}_i x\otimes\overline{\pi}'_i x)+\mathcal{D}^*_{p-k}\phi(V(\overline{\pi})x\otimes V(\overline{\pi})x)\\ &= \sum_i (\mathcal{D}^*_{p-k+1}\delta^{\otimes}+\delta\mathcal{D}^*_{p-k+1})\phi(\overline{\pi}_i x\otimes\overline{\pi}'_i x)+\mathcal{D}^*_{p-k}\phi(V(\overline{\pi})x\otimes V(\overline{\pi})x)\end{aligned}$$

where we have used (1.101) and the duals of (1.127). But the terms involving $\delta^{\otimes}$ vanish because x is a cocycle, and coboundaries represent zero in $\mathrm{Ext}^{*,*}_{\Xi}(M,N)$. So $\overline{\pi}Sq^k\alpha$ is represented by the cocycle $\mathcal{D}^*_{p-k}\phi(V(\overline{\pi})x\otimes V(\overline{\pi})x)$, which in turn represents $Sq^kV(\overline{\pi})\alpha$ in $\mathrm{Ext}^{*,*}_{\Xi}(M,N)$, so we are done. □

2. Left Action by an Algebra with Coproducts

The results of this section will be needed only for our description of the vertical Steenrod operations on the change-of-rings spectral sequence in Chapter 5. Our purpose is to point out that all the definitions and results of the previous section go through if, instead of assuming Π a bialgebra, we take it to be an algebra with coproducts. We fix an algebra with coproducts Π, a right Π-module M and a left Π-module N. We will also assume given a right Π-algebra Ξ; and we will assume that M and N are right Ξ-modules. The main result of this section is Proposition 3.28, which says that if M is a right $\Pi-\Xi$ module in the sense of Definition 1.57, and if the actions of Π and Ξ are "compatible" in the sense of Definition 3.25 below, then formula (3.1) defines an action of $\overline{\Pi}$ upon $\mathrm{Hom}_{\Xi}(M,N)$. We will then extend our definition of $\overline{\Pi}$-action on $\mathrm{Hom}_{\Xi}(M,N)$ to an action of $\overline{\Pi}$ on $\mathrm{Ext}^{p,*,*}_{\Xi}(M,N)$ for all $p\geq 0$. This is accomplished in Definition 3.31, with the use of relative homological algebra. Finally, with appropriate extra structures present on M,N, and Ξ, one can use (1.126) to define Steenrod operations on $\mathrm{Ext}^{*,*,*}_{\Xi}(M,N)$. We will discuss the relationship between these operations, and our newly defined action of $\overline{\Pi}$. The main result in this direction is Proposition 3.32.

We assume our ground field l has characteristic $p>0$, and we remind the reader of our definitions and conventions concerning algebras with coproducts, as given in Section 6 of Chapter 1. Thus, if Π is an algebra with coproducts graded on an abelian group $\mathbb{T}\times\mathbb{Z}$ we write $\overline{\Pi}$ for its negative, as in Definition 1.70. We assume N a left Π-module as in Definition 1.50, and M a right Π-module as in Definition 1.49, both graded on the same abelian group $\mathbb{T}\times\mathbb{S}$. Here $\mathbb{T}$ is the "additive" degree and $\mathbb{S}$ the "multiplicative", as in the just-cited definitions. Then $\mathrm{Hom}(M,N)$ is also graded on $\mathbb{T}\times\mathbb{S}$, as in (1.3a). We check that if $\overline{\pi}\in\overline{\Pi}_{k,v}$ and $f\in\mathrm{Hom}^{t,s}(M,N)$, then $\overline{\pi}\cdot f$ as defined by (3.1) lies in $\mathrm{Hom}^{k+t,p^v s}(M,N)$, as required by our definition of a left action by an algebra with coproducts $\overline{\Pi}$. In carrying out this check one bears in mind that the elements π_i and π'_i in (3.1) lie in degrees $\Pi_{i,v},\Pi_{i',v}$ respectively, for which $i+i'=-k$. By analogy with Proposition 3.3 we have:

PROPOSITION 3.22. *Let Π be an algebra with coproducts graded on $\mathbb{T}\times\mathbb{Z}$, and M a right Π-module and N a left Π-module, both graded on the same group $\mathbb{T}\times\mathbb{S}$. Then under definition* (3.1), $\mathrm{Hom}(M,N)$ *becomes a left $\overline{\Pi}$ module.*

The verification that $(\sigma\tau)\cdot f = \sigma\cdot(\tau\cdot f)$ for all $\sigma, \tau \in \overline{\Pi}$ is formally identical to the proof of Proposition 3.3. One uses the fact that the product $\mu : \Pi_{*,u} \otimes \Pi_{*,v} \to \Pi_{*,u+v}$ is a morphism of coalgebras for all $(u, v) \in \mathbb{Z}\times\mathbb{Z}$, as well as our assumptions, in Definitions 1.49 and 1.50, that actions of an algebra with coproducts on its left and right modules are parity-preserving in the sense of Definition 1.6.

Naturality of the action we have just constructed is expressed by:

PROPOSITION 3.23. *Let Π be an algebra with coproducts graded on $\mathbb{T}\times\mathbb{Z}$. Suppose M, M' a pair of right Π-modules graded on the same group $\mathbb{T}\times\mathbb{S}$; and suppose N, N' a pair of left Π modules also graded on $\mathbb{T}\times\mathbb{S}$. Suppose $g : M' \to M$ is a homomorphism of right Π-modules. Suppose $h : N \to N'$ is a homomorphism of left Π-modules. Then the induced map*

$$< g, h >: \mathrm{Hom}(M, N) \to \mathrm{Hom}(M', N')$$

is a homomorphism of left $\overline{\Pi}$-modules.

Now we need an analogue of Proposition 3.6 in which the objects Π and Π' are no longer bialgebras, but algebras with coproducts graded on $\mathbb{T}\times\mathbb{Z}$, and the modules M, N and $M'N'$ are assumed modules over Π and Π', respectively, all graded on the same group $\mathbb{T}\times\mathbb{S}$. We recall that the cross product $\Pi\times\Pi'$ of algebras with coproducts has been defined by (1.33); and that by Definition 1.52, $M\otimes M'$ and $N\otimes N'$ become right and left modules, respectively, over $\Pi\times\Pi'$. So by Proposition 3.22 the right-hand side of (1.7a) becomes a left module over the negative of $\Pi\times\Pi'$; i.e., over the algebra with coproducts $\overline{\Pi}\times\overline{\Pi}'$. On the other hand, we have from Proposition 3.22 that $\mathrm{Hom}(M, N)$ is a left $\overline{\Pi}$-module and $\mathrm{Hom}(M', N')$ is a left $\overline{\Pi'}$-module, so the left-hand side of (1.7a) also is a left module over $\overline{\Pi}\times\overline{\Pi}'$.

PROPOSITION 3.24. *Under the hypotheses above, the pairing of* (1.7a)

$$\nu : \mathrm{Hom}(M, N)\otimes\mathrm{Hom}(M', N',) \to \mathrm{Hom}(M\otimes M', N\otimes N')$$

is a homomorphism of $\overline{\Pi}\times\overline{\Pi}'$-modules.

This is easily checked: one uses the hypothesis that actions of algebras with coproducts on their modules are parity-preserving in the sense of Definition 1.6.

Next we need the analogue of Proposition 3.14 for algebras with coproducts Π. The notion of a right algebra over an algebra with coproducts is already in Remark 1.54. We assume Ξ a right algebra over Π that is graded on the same group $\mathbb{T}\times\mathbb{S}$ as the modules M and N above. Then the notion of M being a right $\Pi-\Xi$ module is already in Definition 1.57. So we only need to find the right definition of a module N with "compatible" actions of Π on the left and Ξ on the right. The diagram (3.3) will not do: the action of Π upon N is in the present context not a homomorphism from $\Pi\otimes N$ to N. Instead, we use the definition (3.4) in terms of elements:

DEFINITION 3.25. Let Π be an algebra with coproducts graded on $\mathbb{T}\times\mathbb{Z}$, and Ξ a right Π-algebra graded on $\mathbb{T}\times\mathbb{S}$. By a module with compatible actions of Π on the left and Ξ on the right we mean a left Π-module N graded on $\mathbb{T}\times\mathbb{S}$, that is also a right Ξ-module, in such a way that (3.4) holds for all $n \in N$, $\xi \in \Xi$, $\pi \in \Pi$. Here the expression $\xi\cdot\pi_i'$ refers to the right action of Π upon Ξ.

It is a nice review of our definitions to check that (3.4) is compatible with gradings. That is, if we start with $\pi \in \Pi_{k,v}, n \in N^{t,s}, \xi \in \Xi_{t',s'}$ for $k, t, t' \in \mathbb{T}$, $s, s' \in \mathbb{S}$ and $v \in \mathbb{Z}$, then both sides of (3.4) lie in $N^{k+t+t',p^v s+s'}$.

REMARK 3.26. Propositions 3.10 through 3.13 all have obvious analogues in which the bialgebras Π, Π', and Δ are replaced by algebras with coproducts; and the tensor product of bialgebras wherever it occurs is replaced by the cross product of algebras with coproducts. The proofs of all of these are easy. The version of Proposition 3.12 in which Π and Π' are algebras with coproducts is proved by direct computation from (3.4).

In particular we record the analogue of Proposition 3.13:

PROPOSITION 3.27. *Let Π be an algebra with coproducts and Ξ a right bialgebra over Π. Suppose N and N' a pair of modules, each with compatible actions of Π on the left and Ξ on the right. Then the diagonal actions of Π and Ξ on $N \otimes N'$ are also compatible.*

We have introduced the notion of compatible actions so that we can state and prove the following analogue of Proposition 3.14.

PROPOSITION 3.28. *Let Π be an algebra with coproducts and let Ξ be a right Π-algebra. Let M and N be a pair of right Ξ-modules. Suppose given actions of Π on the right of M, and on the left of N, in such a way that:*

1. *M is a right $\Pi - \Xi$ module;*
2. *the actions of Π and Ξ on N are compatible in the sense of Definition 3.25.*

Let $\mathrm{Hom}(M, N)$ is regarded as a left $\overline{\Pi}$-module, as in (3.1). Then $\mathrm{Hom}_\Xi(M, N)$ is a sub-$\overline{\Pi}$ module. Thus, (3.1) defines a action:

$$\overline{\Pi} \otimes \mathrm{Hom}_\Xi(M, N) \to \mathrm{Hom}_\Xi(M, N).$$

This action is natural in the sense that if $g : M \to M'$ a homomorphism of right $\Pi - \Xi$ modules, and if $h : N' \to N$ is a homomorphism of modules with compatible actions of Π on the left and Ξ on the right, then the induced mapping:

$$\langle g, h\rangle : \mathrm{Hom}(M', N') \to \mathrm{Hom}(M, N)$$

is a homomorphism of left $\overline{\Pi}$-modules.

The proof is formally identical with the proof of Proposition 3.14. Naturality is an immediate consequence of Propostion 3.23.

We need an analogue of Proposition 3.17. So we suppose we have two algebras with coproducts Π and Π'; we suppose Ξ a right Π-algebra and Ξ' a right Π'-algebra. Let M, N be a pair of right modules over Ξ, and M', N' a pair of right modules over Ξ'. We assume given actions of Π on the right of M, of Π' on the right of M'; and actions of Π on the left of N, of Π' on the left of N', in such a way that:

1. M is a right $\Pi - \Xi$ module and M' is a right $\Pi' - \Xi'$ module;
2. the actions of Π and Ξ on N are compatible, and the actions of Π' and Ξ' on N' are compatible.

Then from Propostion 1.60 we have that $M \otimes M'$ is a right $\Pi \times \Pi' - \Xi \otimes \Xi'$ module. From Proposition 3.12...modified so that the bialgebras are replaced by algebras with coproducts ... we have that the actions of $\Pi \times \Pi'$ and $\Xi \otimes \Xi'$ on $N \otimes N'$ are compatible. Thus, Proposition 3.28 gives us not only actions of $\overline{\Pi}$ on $\mathrm{Hom}_\Xi(M, N)$ and of $\overline{\Pi}'$ upon $\mathrm{Hom}_{\Xi'}(M', N')$; but it also gives us an action of $\overline{\Pi} \times \overline{\Pi}'$ upon $\mathrm{Hom}_{\Xi \otimes \Xi'}(M \otimes M', N \otimes N')$.

PROPOSITION 3.29. *Under the hypotheses above, the pairing of* (1.23a)*:*

$$(3.15) \qquad \nu : \mathrm{Hom}_{\Xi}(M,N) \otimes \mathrm{Hom}_{\Xi'}(M',N',) \to \mathrm{Hom}_{\Xi\otimes\Xi'}(M \otimes M', N \otimes N')$$

is a homomorphism of $\overline{\Pi} \times \overline{\Pi}'$*-modules.*

This follows at once from Proposition 3.24.

Finally we need an analogue of Propostion 3.19. So we suppose Π is an algebra with coproducts, and Ξ is a right Π-bialgebra, as in Remark 1.54. Suppose M, N a pair of right Ξ-modules having actions of Π satisfying the hypotheses of Proposition 3.28. Let M', N' be another such pair. We then give $M \otimes M'$ and $N \otimes N'$ the diagonal actions of both Ξ and Π. Then by Proposition 1.61, $M \otimes M'$ is a $\Pi - \Xi$ module, and by Proposition 3.27, the actions of Π and Ξ on $N \otimes N'$ are compatible. Thus Proposition 3.28 gives $\overline{\Pi}$ actions on all of $\mathrm{Hom}_{\Xi}(M,N)$, $\mathrm{Hom}_{\Xi}(M',N')$, and $\mathrm{Hom}_{\Xi}(M \otimes M', N \otimes N')$.

PROPOSITION 3.30. *Suppose* Π *is an algebra with coproducts, and* Ξ *is a right* Π*-bialgebra. Let* M, N *be a pair of right modules over* Ξ*, having actions of* Π *satisfying the hypothesis of Proposition 3.28. Let* M', N' *be another such pair of right modules over* Ξ*. Then the pairing:*

$$(3.16) \qquad \rho : \mathrm{Hom}_{\Xi}(M,N) \otimes \mathrm{Hom}_{\Xi}(M',N') \to \mathrm{Hom}_{\Xi}(M \otimes M', N \otimes N')$$

of (1.30) *is a homomorphism of* $\overline{\Pi}$*-modules, with* $\overline{\Pi}$ *acting diagonally on the domain of* ρ*.*

This follows from Proposition 3.29 in the same way that Proposition 3.19 follows from Proposition 3.17. The key point is that the coproduct $\psi : \Xi \to \Xi \otimes \Xi$ is a homomorphism of Π-algebras.

Now we need to extend the action of $\overline{\Pi}$ upon $\mathrm{Hom}_{\Xi}(M,N)$ that we have constructed in Proposition 3.28 to an action of $\overline{\Pi}$ upon $\mathrm{Ext}^p_{\Xi}(M,N)$ for each $p \geq 0$. We have been assuming Ξ, M, and N all graded on the same group $\mathbb{T} \times \mathbb{S}$, with $\mathbb{T}$ giving the additive degree, and $\mathbb{S}$ the multiplicative. So for each $p \geq 0$, $\mathrm{Ext}^p_{\Xi}(M,N)$ is also graded on this group ... $\mathrm{Ext}^p_{\Xi}(M,N) = \{\mathrm{Ext}^{p,t,s}_{\Xi}(M,N) \mid t \in \mathbb{T},\ s \in \mathbb{S}\}$. We recall from Chapter 1, Remark 1.64, the notion of a relatively projective $\Pi - \Xi$ module; also from Definition 1.68 the notion of a resolution $\mathcal{P}(M) \to M \to 0$ of an arbitrary $\Pi - \Xi$ module M by relative projectives. Remark 1.69 assures that such a resolution always exists. By definition, a resolution of a $\Pi - \Xi$ module M by relative projectives is also a projective resolution of M as a Ξ-module, in the ordinary sense. So for any right Ξ-module N we have

$$(3.17) \qquad \mathrm{Ext}^{p,*,*}_{\Xi}(M,N) = H^{p,*,*}\,\mathrm{Hom}_{\Xi}(\mathcal{P}(M),N),$$

an isomorphism of $\mathbb{T} \times \mathbb{S}$-graded vector spaces for each $p \geq 0$. But suppose N also has a left Π-action that is compatible with the action of Ξ, in the sense of Definition 3.25. Then by Proposition 3.28, $\mathrm{Hom}_{\Xi}(\mathcal{P}(M),N)$ becomes a cochain complex over the category of left $\overline{\Pi}$-modules. So (3.17) defines a left $\overline{\Pi}$-action on $\mathrm{Ext}^{p,*,*}_{\Xi}(M,N)$, for each $p \geq 0$. By the uniqueness of the projective resolution up to chain homotopy equivalence of $\Pi - \Xi$ complexes (Remark 1.69), the action of $\overline{\Pi}$ is well defined. We summarize:

DEFINITION 3.31. Let Π be an algebra with coproducts , Ξ a right Π-algebra, M a right $\Pi - \Xi$ module, and N a module with compatible actions of Π on the left and Ξ on the right. Let $\mathcal{P}(M)$ be any resolution of M by relative projectives in the category of $\Pi - \Xi$ modules. Then $\mathrm{Hom}_\Xi(\mathcal{P}(M), N)$ is a cochain complex over the category of left $\overline{\Pi}$-modules, so the equation (3.17) defines $\mathrm{Ext}_\Xi^{p,*,*}(M,N)$ as a left $\overline{\Pi}$-module, for each $p \geq 0$. The action of $\overline{\Pi}$ is independent of the choice of resolution.

Now, as in the previous section, we fix the ground field to be $l = \mathbb{Z}/2$, and we consider the situation in which products and Steenrod operations are defined on $\mathrm{Ext}_\Xi^{*,*,*}(M,N)$. We ask how these are related to the action of $\overline{\Pi}$ that we constructed in Definition 3.31. In order to define products and Steenrod operations we need, in addtion to the hypotheses of Proposition 3.28, that Ξ be a Hopf algebra, M a coalgebra, and N an algebra; and that these new structures be compatible with the actions of Π upon $\Xi, M,$ and N; and of Ξ upon M and N. In this connection we remind the reader of the Definitions in Remark 1.54.

In the following proposition we will regard $\mathrm{Ext}_\Xi^{*,*,*}(M,N)$ as an algebra graded on $\mathbb{Z} \times \mathbb{T} \times \mathbb{S}$, where the first factor gives homological degree. We will also regard $\overline{\Pi}$ as an algebra with coproducts graded on $(\mathbb{Z} \times \mathbb{T}) \times \mathbb{Z}$. It is concentrated in the degrees $(0 \times \mathbb{T}) \times \mathbb{Z}$, with the factor in parentheses giving the additive degree, and the last factor of $\mathbb{Z}$ the multiplicative degree. With these conventions, we can regard equation (3.9) as defining $\mathrm{Ext}_\Xi^{*,*,*}(M,N)$ as a left module over the algebra with coproducts $\overline{\Pi}$.

PROPOSITION 3.32. *Let Π be an algebra with coproducts , and Ξ a right Π-Hopf algebra. Suppose M a right $\Pi - \Xi$ module, and N a module with compatible actions of Π on the left and Ξ on the right. Suppose in addition that M is a coalgebra and N a commutative algebra, in such a way that:*

1. *M is a Ξ-coalgebra, and a Π-coalgebra;*
2. *N is a Ξ-algebra and a Π-algebra.*

With the action of $\overline{\Pi}$ on $\mathrm{Ext}_\Xi^{*,*,*}(M,N)$ defined by (3.17), and the product on $\mathrm{Ext}_\Xi^{*,*,*}(M,N)$ as in (1.125), then $\mathrm{Ext}_\Xi^{*,*,*}(M,N)$ becomes a commutative $\overline{\Pi}$-algebra. Further, with the Steenrod operations on $\mathrm{Ext}_\Xi^{*,*,*}(M,N)$ defined as in (1.126) we have

$$\overline{\pi} Sq^k \alpha = Sq^k[V(\overline{\pi})\alpha] \tag{3.18}$$

for all $\overline{\pi} \in \overline{\Pi}$, all $\alpha \in \mathrm{Ext}_\Xi^{*,*,*}(M,N)$ and all $k \geq 0$. Here V is the Verschiebung of (1.17).

The proof of this Proposition is formally identical with the proof of Proposition 3.21. As in that earlier proof we construct a resolution $\mathcal{P}(M) \to M \to 0$ of the right $\Pi - \Xi$ module M by relative projectives, and a "homological cup-k product", of which each mapping $\mathcal{D}_k$ is linear over both Ξ and Π. The $\overline{\Pi}$-linearity of the cochain mapping ϕ of (3.12) in each homological degree follows in the present context from Proposition 3.30, by way of diagram (3.13). Of course the Π-linearity of the product $\mu : N \otimes N \to N$ is essential to the argument. Now we see that the cochain operation m of (3.14) is $\overline{\Pi}$-linear in each homological degree. We have as a consequence that $\mathrm{Ext}_\Xi^{*,*,*}(M,N)$ is a $\overline{\Pi}$-algebra; this is the first part of our theorem. As for the second, the proof of (3.18) is the same as the proof of (3.10). One need only keep in mind that $\pi \in \Pi$ lies in a particular multiplicative degree; say, $\pi \in \Pi_{*,v}$; so that the expansion (1.17) and the Verschiebung $V(\overline{\pi})$ are calculated in the coalgebra $\Pi_{*,v}$.

3. Right Action by a Hopf algebra

In this section we again fix two right modules M, N over an algebra Ξ, and find a set of conditions under which a bialgebra $\overline{\Pi}$ can act on the group of Ξ-linear homomorphisms from M to N. In the previous cases Π was acting on the right of M and on the left on N. Now we will assume Π acting on the right of both. Then in order to define an action of $\overline{\Pi}$ on $\mathrm{Hom}_\Xi(M,N)$ we will have to assume that Π has an antipode; i.e., that Π is a Hopf algebra. Also we will formulate our theory so that $\overline{\Pi}$ acts from the right of $\mathrm{Hom}_\Xi(M,N)$, rather than from the left as in the previous cases. This version will be useful in Chapters 4 and 5, when we set up the change-of-rings spectral sequence for the cohomology of an extension of Hopf algebras, and describe the Steenrod operations at the E_2-level.

Clearly if N is a right module over a Hopf algebra Π, one could regard it also as a left module, the left action being defined in terms of the right according to the formula

$$\pi n = (-1)^{|\pi||n|} n\chi(\pi)$$

for all $\pi \in \Pi$, $n \in N$. So one could apply the theory of Section 1 to obtain a left action of $\overline{\Pi}$ on $\mathrm{Hom}_\Xi(M,N)$, and then translate this back into a right action according to:

$$f\overline{\pi} = (-1)^{|\pi||f|}\chi(\overline{\pi})f$$

for all $\pi \in \Pi$, $f \in \mathrm{Hom}_\Xi(M,N)$. The main definitions and theorems in this section can be obtained from the definitions and theorems of Section 1 by these purely formal operations, so no further proofs are needed. We have nevertheless written out all the definitions and theorem statements, for ease of application in Chapters 4 and 5. We have also provided some indications of proof for the main results, that are independent of Section 1.

The main result of this section is Proposition 3.42, which shows that if M is a right $\Pi - \Xi$ module, and if the right actions of Π and Ξ on N are "compatible" in the sense of Definition 3.37, then (3.19) defines a right action of $\overline{\Pi}$ on $\mathrm{Hom}_\Xi(M,N)$. Then in Definition 3.45 we use relative homological algebra to extend our action of $\overline{\Pi}$ on $\mathrm{Hom}_\Xi(M,N)$ to an action of $\overline{\Pi}$ on $\mathrm{Ext}_\Xi^{q,*}(M,N)$, for each $q \geq 0$. Finally, with appropriate extra structures present on M, N, and Ξ, one can use (1.126) to define Steenrod operations on $\mathrm{Ext}_\Xi^{*,*}(M,N)$. We will discuss the relationship between these operations, and our newly defined action of $\overline{\Pi}$. The main result in this direction is Proposition 3.47.

Fix a Hopf algebra Π graded on an abelian group $\mathbb{S}$, with antipode $\chi : \Pi \to \Pi$. Suppose M and N are a pair of right Π-modules.

DEFINITION 3.33. Given $f \in \mathrm{Hom}(M,N)$ and $\pi \in \Pi$, define $f\cdot\overline{\pi} \in \mathrm{Hom}(M,N)$ by setting

$$(f \cdot \overline{\pi})(m) = (-1)^{|\pi||m|}\sum_i f\big(m(\chi\pi_i)\big)\pi_i' \tag{3.19}$$

for all $m \in M$.

Of course, Remark 3.2 is relevant here as well.

PROPOSITION 3.34. *Under the definition* (3.19), $\mathrm{Hom}(M,N)$ *becomes a right* $\overline{\Pi}$-*module.*

In fact, the reader will check that the mapping $\mathrm{Hom}(M, N) \otimes \overline{\Pi} \to \mathrm{Hom}(M, N)$ defined by (3.19) preserves degree. The verification that $f \cdot (\sigma\tau) = (f \cdot \sigma) \cdot \tau$ for all $\sigma, \tau \in \overline{\Pi}$ is a straightforward computation that uses the fact that χ is an antiautomorphism of algebras.

The action of $\overline{\Pi}$ that we have just constructed is natural, in the following sense:

PROPOSITION 3.35. *Suppose $g : M' \to M$ is a homomorphism of right Π-modules homogeneous of any degree. Suppose that $h : N \to N'$ is a homomorphism of right Π-modules homogeneous of any degree. Then the induced map*

$$\langle g, h\rangle : \mathrm{Hom}(M, N) \to \mathrm{Hom}(M', N')$$

is a homomorphism of right $\overline{\Pi}$-modules.

We will need the analogue of Proposition 3.6. So we suppose given a pair of Hopf algebras Π, Π'; a pair of right Π-modules M, N, and a pair of right Π'-modules M', N'. Then from Proposition 3.34 we have actions of $\overline{\Pi}$ on $\mathrm{Hom}(M, N)$, of $\overline{\Pi'}$ on $\mathrm{Hom}(M', N')$, and of $\overline{\Pi} \otimes \overline{\Pi'}$ on $\mathrm{Hom}(M \otimes M', N \otimes N')$.

PROPOSITION 3.36. *Under the hypotheses above, the pairing of* (1.7a)*:*

$$\nu : \mathrm{Hom}(M, N) \otimes \mathrm{Hom}(M', N',) \to \mathrm{Hom}(M \otimes M', N \otimes N') \tag{3.20}$$

is a homomorphism of $\overline{\Pi} \otimes \overline{\Pi'}$-modules.

This is an easy consequence of the definitions; mainly one must keep track of the signs.

Next we need the analogue of Definition 3.8.

DEFINITION 3.37. Let Π be a Hopf algebra, and Ξ a right Π-algebra. Suppose N a right Π-module with action $N \otimes \Pi \xrightarrow{\sigma_N} N$, that is also a right Ξ-module with action $N \otimes \Xi \xrightarrow{\sigma'_N} N$. We say the actions of Π and Ξ are compatible if the following diagram commutes:

(3.21)

$$\begin{array}{ccccc}
N \otimes \Pi \otimes \Xi & \xrightarrow{\sigma_N \otimes \Xi} & N \otimes \Xi & \xrightarrow{\sigma'_N} & N \\
\downarrow{\scriptstyle (1,3,2)} & & & & \uparrow{\scriptstyle \sigma_N} \\
N \otimes \Xi \otimes \Pi & & & & N \otimes \Pi \\
\downarrow{\scriptstyle N \otimes \Xi \otimes \psi} & & & & \uparrow{\scriptstyle \sigma'_N \otimes \Pi} \\
N \otimes \Xi \otimes \Pi \otimes \Pi & \xrightarrow[N \otimes \Xi \otimes \chi \otimes \Pi]{} & N \otimes \Xi \otimes \Pi \otimes \Pi & \xrightarrow[N \otimes \sigma_\Xi \otimes \Pi]{} & N \otimes \Xi \otimes \Pi.
\end{array}$$

We will speak of a module N satisfying this definition as "a module with compatible right actions of Π and Ξ".

We will find it useful to express the commutativity of (3.21) in terms of elements:

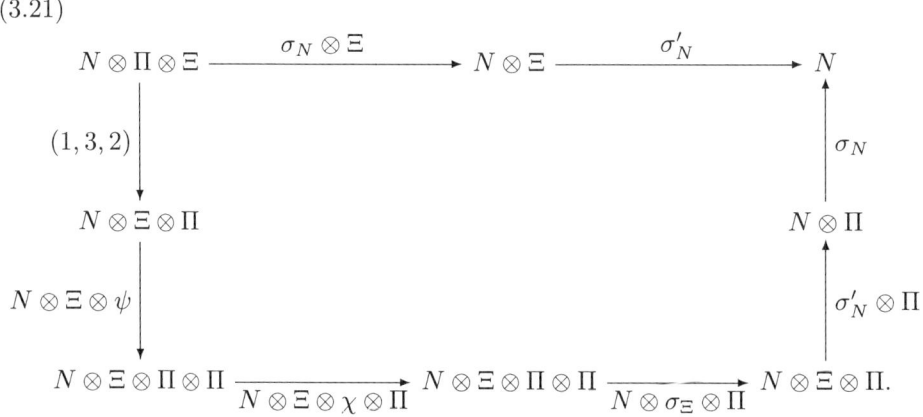

$$(n\pi)\xi = (-1)^{|\xi||\pi|} \sum_i \big(n(\xi \cdot \chi\pi_i)\big)\pi_i' \tag{3.22}$$

for all $n \in N, \pi \in \Pi, \xi \in \Xi$. Here the expression $\xi \cdot \chi\pi_i$ refers to the right action of Π on Ξ.

EXAMPLE 3.38. Suppose C a group and A a normal subgroup. Write $\Pi = l(C)$ and $\Xi = l(A)$ for the corresponding group rings over our ground field l. In the usual way we regard these are Hopf algebras concentrated in degree 0; the coproducts are $\psi(c) = c \otimes c$, $\psi(a) = a \otimes a$, for all $c \in C,\ a \in A$. Define a right action of Π on Ξ by setting $a \cdot c = c^{-1}ac$ for all $c \in C,\ a \in A$. Then Ξ is a right Π-algebra. Now if N is a right Π-module we regard it also as a Ξ-module by restriction. It is easy to check that (3.22) is satisfied, so N becomes a module with compatible right actions of Π and Ξ.

In Chapter 4, Proposition 4.27 we will generalize this example, replacing $l(C)$ by an arbitrary cocommutative Hopf algebra, and $l(A)$ by a normal sub-Hopf algebra.

If Π is a Hopf algebra and Ξ a right Π-algebra, then by a sub Π-algebra of Ξ we mean a sub-algebra of Ξ that is also a sub Π-module. The following proposition will be useful in Chapter 4. Its proof is clear.

PROPOSITION 3.39. *Let Π be a Hopf algebra, Ξ a right Π-algebra, and Ξ' a sub Π-algebra of Ξ. Let N be a module with compatible right actions of Π and Ξ. Then if N is regarded as a right Ξ' module by restriction of the action of Ξ, the right actions of Π and Ξ' are also compatible.*

REMARK 3.40. Propositions 3.10 through 3.13 all have obvious analogues in the present context. That is, we assume the bialgebras Π, Π', Δ are Hopf algebras. Objects like Ξ are still right Π-algebras, but we replace modules with compatible actions of Π on the left and Ξ on the right by modules with compatible right actions of Π and Ξ. It is easy to show that Propositions 3.10 through 3.13 remain true with these substitutions.

In particular we record the analogue of Proposition 3.13:

PROPOSITION 3.41. *Let Π be a Hopf algebra and Ξ a right bialgebra over Π. Suppose N and N' a pair of modules, each with compatible right actions of Π and Ξ. Then the diagonal actions of Π and Ξ on $N \otimes N'$ are compatible.*

We have introduced the notion of compatible right actions so that we can state and prove the following analogue of Proposition 3.14.

PROPOSITION 3.42. *Let Π be a Hopf algebra, and Ξ a right Π-algebra. Suppose M, N a pair of right Ξ-modules that are also right Π-modules, in such a way that:*

1. *M is a right $\Pi - \Xi$ module in the sense of Definition 1.32;*
2. *the actions of Π and Ξ on N are compatible in the sense of Definition 3.37.*

Let $\mathrm{Hom}(M, N)$ *be regarded as a right $\overline{\Pi}$-module, as in Definition 3.33. Then* $\mathrm{Hom}_\Xi(M, N)$ *is a sub-Π module. Thus,* (3.19) *defines an action:*

$$\mathrm{Hom}_\Xi(M, N) \otimes \overline{\Pi} \to \mathrm{Hom}_\Xi(M, N). \tag{3.23}$$

This action is natural in the sense that if $g : M' \to M$ is a homomorphism of $\Pi - \Xi$ modules, and if $h : N \to N'$ is a homomorphism of modules with compatible right *actions of* Π *and* Ξ*, then the induced mapping:*

$$\langle g,h\rangle : \mathrm{Hom}_\Xi(M,N) \to \mathrm{Hom}_\Xi(M',N')$$

is a homomorphism of right $\overline{\Pi}$*-modules.*

The proof that $(f \cdot \overline{\pi})(m\xi)) = [(f \cdot \overline{\pi})(m)]\xi$ is a routine computation with reference to (3.19) and (3.22). As in the proof of Proposition 3.14 we need the cocommutativity of Π. Naturality is already in Proposition 3.35.

Suppose we have a chain complex C over the category of right $\Pi - \Xi$ modules, and suppose given a module N with compatible right actions of Π and Ξ. Applying the above proposition in each homological degree of the dual, cochain complex $\mathrm{Hom}_\Xi(C,N)$, we find we have a cochain complex over the category of right $\overline{\Pi}$-modules. Thus, $H^{q,*}\,\mathrm{Hom}_\Xi(C,N)$ acquires the structure of a right $\overline{\Pi}$-module for each integer q. Then we have:

PROPOSITION 3.43. *Let* Π *be a Hopf algebra and* Ξ *a right* Π*-algebra. Suppose given a short-exact sequence of chain complexes over the category of right* $\Pi - \Xi$ *modules*

$$0 \to C' \to C \to C'' \to 0,$$

where C''_q *is projective over* Ξ *in each homological degree* q*. If* N *is any module with compatible right actions of* Π *and* Ξ*, let*

$$0 \to \mathrm{Hom}_\Xi(C'',N) \to \mathrm{Hom}_\Xi(C,N) \to \mathrm{Hom}_\Xi(C',N) \to 0$$

be the associated short-exact sequence of cochain complexes, and let

$$\Delta : H^{q,*}\,\mathrm{Hom}_\Xi(C',N) \to H^{q+1,*}\,\mathrm{Hom}_\Xi(C'',N)$$

be the associated connecting morphism. Then Δ *is a homomorphism of* $\overline{\Pi}$*-modules in each homological degree* q*.*

This is a straightforward consequence of the definitions. Of course, a similar theorem could have been proved in the contexts of the first two sections of this chapter; but the present context is the only one in which we will need such a result. We will use Proposition 3.43 in Chapter 4.

We study the relation of the $\overline{\Pi}$-action constructed in Proposition 3.42 to the pairing ρ of (1.29a). We suppose Π a Hopf algebra, and that Ξ is a right Π-bialgebra. Let M, N be a pair of right modules over Ξ, having actions of Π satisfying the hypothesis of Proposition 3.42. Let M', N' be another such pair. We then give $M \otimes M'$ and $N \otimes N'$ the diagonal actions of both Ξ and Π. Then we have from from Propositions 1.37 and 3.41 that $M \otimes M'$ is a $\Pi - \Xi$ module, and that the actions of Π and Ξ on $N \otimes N'$ are compatible. Thus Proposition 3.42 gives $\overline{\Pi}$ actions not only on $\mathrm{Hom}_\Xi(M,N)$ and $\mathrm{Hom}_\Xi(M',N')$, but also on $\mathrm{Hom}_\Xi(M \otimes M', N \otimes N')$. Under these conditions we have:

PROPOSITION 3.44. *The pairing*

$$\rho : \mathrm{Hom}_\Xi(M,N) \otimes \mathrm{Hom}_\Xi(M',N') \to \mathrm{Hom}_\Xi(M \otimes M', N \otimes N')$$

is a homomorphism of right $\overline{\Pi}$*-modules, if* $\overline{\Pi}$ *acts diagonally on the domain of* ρ*.*

This follows from Proposition 3.36 in the same way the Proposition 3.19 follows from Proposition 3.6. Next we extend the action of $\overline{\Pi}$ on $\mathrm{Hom}_\Xi(M,N)$ that we constructed in Proposition 3.42 to an action of $\overline{\Pi}$ on $\mathrm{Ext}^q_\Xi(M,N)$ for all $q \geq 0$. The procedure is the same as the one we followed in Section 1. Given a Hopf algebra Π and a right Π-algebra Ξ, we have relatively projective right $\Pi - \Xi$ modules as in Definition 1.41. Given a $\Pi - \Xi$ module M, one can build a resolution $\mathcal{P}(M)$ of M by relative projectives in the sense of 1.45. Then for any right Ξ-module N we have

$$\operatorname{Ext}_{\Xi}^{q,*}(M,N) = H^{q,*}\operatorname{Hom}_{\Xi}(\mathcal{P}(M),N), \tag{3.24}$$

an isomorphism of $\mathbb{S}$-graded vector spaces for each $q \geq 0$. But suppose N also has a right Π-action that is compatible with the action of Ξ, in the sense of Definition 3.37. Then by Proposition 3.42, $\operatorname{Hom}_{\Xi}(\mathcal{P}(M),N)$ becomes a cochain complex over the category of right $\overline{\Pi}$-modules. So (3.24) defines a right $\overline{\Pi}$-action on $\operatorname{Ext}_{\Xi}^{q,*}(M,N)$, for each $q \geq 0$. By the uniqueness of the projective resolution up to chain homotopy equivalence of $\Pi - \Xi$ complexes, the action of $\overline{\Pi}$ is well defined. We summarize:

DEFINITION 3.45. Let Π be a Hopf algebra, Ξ a right Π-algebra, M a right $\Pi - \Xi$ module, and N a module with compatible right actions of Π and Ξ. Let $\mathcal{P}(M) \to M \to 0$ be any resolution of M by relative projectives in the category of $\Pi - \Xi$ modules. Then (3.24) defines a right action of $\overline{\Pi}$ on $\operatorname{Ext}_{\Xi}^{q,*}(M,N)$, for each $q \geq 0$, that is independent of the choice of resolution.

PROPOSITION 3.46. *Let Π be a Hopf algebra, Ξ a right Π-algebra, and N a module with compatible right actions of Π and Ξ. If:*

$$0 \to M' \to M \to M'' \to 0$$

is any short-exact sequence of right $\Pi - \Xi$ modules, the connecting morphism:

$$\Delta : \operatorname{Ext}_{\Xi}^{q,*}(M',N) \to \operatorname{Ext}_{\Xi}^{q+1,*}(M'',N)$$

is a homomorphism of $\overline{\Pi}$-modules, for each $q \geq 0$.

PROOF. The classical arguments of homological algebra, in conjunction with Proposition 1.44, enable us to construct a short-exact sequence of chain complexes of $\Pi - \Xi$ modules:

$$0 \to \mathcal{P}(M') \to \mathcal{P}(M) \to \mathcal{P}(M'') \to 0,$$

where these are resolutions of $M', M, \text{and} M''$, respectively, by relative projectives, in the sense of Definition 1.45. Applying Proposition 3.43 to this sequence completes the proof. □

Now, as in the previous sections, we fix the ground field to be $l = \mathbb{Z}/2$, and consider the situation in which products and Steenrod operations are defined on $\operatorname{Ext}_{\Xi}^{*,*}(M,N)$, as in Proposition 1.120. We ask how these are related to the action of $\overline{\Pi}$ that we constructed in Definition 3.45. In order to define products and Steen rod operations we need, in addtion to the hypotheses of Definition 3.45, that Ξ be a Hopf algebra, that M be a Ξ-coalgebra in the sense of Definition 1.26, and that N be a commutative Ξ-algebra. In order to relate the action of Π to the products and Steenrod operations, we must assume that M is a Π-coalgebra and that N is a Π-algebra.

In the following proposition we will regard $\operatorname{Ext}_{\Xi}^{*,*}(M,N)$ as a vector space graded on $\mathbb{Z} \times \mathbb{S}$, where the first factor gives homological degree. We will also regard Π as a Hopf algebra graded on $\mathbb{Z} \times \mathbb{S}$: it is concentrated in the degrees $0 \times \mathbb{S}$. With these conventions, (3.24) defines $\operatorname{Ext}_{\Xi}^{*,*}(M,N)$ as a right $\overline{\Pi}$-module.

PROPOSITION 3.47. *Let* Π *be a Hopf algebra,* Ξ *a right* Π*-Hopf algebra. Suppose* M *a right* $\Pi - \Xi$ *module, and* N *a module with compatible right actions of* Π *and* Ξ *in the sense of Definition 3.37. Suppose in addition that* M *is a coalgebra and* N *a commutative algebra, in such a way that:*

1. M *is a* Ξ*-coalgebra, and a* Π*-coalgebra;*
2. N *is a* Ξ*-algebra and a* Π*-algebra.*

With the action of $\overline{\Pi}$ *on* $\mathrm{Ext}_{\Xi}^{*,*}(M,N)$ *defined by* (3.24), *and with the product on* $\mathrm{Ext}_{\Xi}^{*,*}(M,N)$ *as in* (1.125), *then* $\mathrm{Ext}_{\Xi}^{*,*}(M,N)$ *becomes a commutative right* $\overline{\Pi}$*-algebra. Further, with the Steenrod operations on* $\mathrm{Ext}_{\Xi}^{*,*}(M,N)$ *defined as in* (1.126) *we have*

$$(Sq^k\alpha)\overline{\pi} = Sq^k(\alpha \cdot V(\overline{\pi}))$$

for all $\overline{\pi} \in \overline{\Pi}$, *all* $\alpha \in \mathrm{Ext}_{\Xi}^{*,*}(M,N)$ *and all* $k \geq 0$. *Here* V *is the Verschiebung of* (1.17).

The proof of this Proposition is formally identical to the proof of Proposition 3.21; just switch the action of $\overline{\Pi}$ from the left of $\mathrm{Ext}_{\Xi}^{*,*}(M,N)$ to the right.

CHAPTER 4

Extensions of Hopf Algebras

Our definition of an extension of Hopf algebras is motivated by the famous Theorems 4.4 and 4.7 of [**51**]. It is the same one that is used in [**30**], [**63**], and [**76**].

Fix an arbitrary ground field l, and assume all modules, algebras, coalgebras, Hopf algebras are vector spaces over l graded on an abelian group $\mathbb{S}$. As has been the case throughout this book, coproducts are assumed commutative.

DEFINITION 4.1. Let

$$\Gamma \xrightarrow{\alpha} \Lambda \xrightarrow{\beta} \Omega \tag{4.1}$$

be a diagram of Hopf algebras. We call this diagram an extension of Hopf algebras if there is a degree-preserving homomorphism of vector spaces $\lambda : \Lambda \to \Omega \otimes \Gamma$ which is simultaneously an isomorphism of right Γ-modules and left Ω-comodules, and which makes the following diagram commute:

$$\begin{array}{ccccc} \Gamma & \xrightarrow{\alpha} & \Lambda & \xrightarrow{\beta} & \Omega. \\ & \searrow^{i_\Gamma} & \downarrow{\scriptstyle \lambda \,\cong} & \nearrow_{p_\Omega} & \\ & & \Omega \otimes \Gamma & & \end{array} \tag{4.2}$$

Here i_Γ is the inclusion $\Gamma = l \otimes \Gamma \xrightarrow{\eta \otimes \Gamma} \Omega \otimes \Gamma$; similarly, p_Ω is the projection $\Omega \otimes \Gamma \xrightarrow{\Omega \otimes \epsilon} \Omega \otimes l = \Omega$.

Given an extension (4.2) our main goal in this chapter is to establish the adjunction isomorphism:

$$\mathrm{Hom}_\Lambda(P \otimes Q, N) \cong \mathrm{Hom}_\Omega(P, \overline{\mathrm{Hom}}^\Gamma(Q, N)). \tag{4.3}$$

Here it is assumed that P is a right Ω-module, that Q and N are right Λ-modules, and that Λ acts on P through the mapping $\beta : \Lambda \to \Omega$. The result appears as Proposition 4.35 below. An important step is to define the action of Ω upon $\overline{\mathrm{Hom}}^\Gamma(Q, N)$. We will do this by first showing how the extension (4.1) defines an action of Λ on Γ by "conjugation"; showing that this action makes Γ into a right Λ-algebra, and then using the theory of Chapter 3, Section 3. We obtain actions of Λ on $\overline{\mathrm{Hom}}^\Gamma(Q, N)$, and on $\overline{\mathrm{Ext}}^\Gamma_{-q,*}(Q, N)$ for each $q \geq 0$. These pass to quotient actions of Ω. In fact we will find two different ways of defining the same action of Λ on $\overline{\mathrm{Ext}}^\Gamma_{-q,*}(Q, N)$. One uses a resolution of Q by relative projectives in the category of $\Lambda - \Gamma$ modules. This is the method used in Chapter 3. Defining the Λ-action in this way allows us to cite all relevant results from that chapter; so that, for example, when we specialize in the next chapter to the case $l = \mathbb{Z}/2$ we will be able to write down the relation of the Λ action on $\overline{\mathrm{Ext}}^{\Gamma}_{-q,*}(Q, N)$ to the action of the Steenrod squares. But we will also develop another method for defining the action of Λ on $\overline{\mathrm{Ext}}^{\Gamma}_{-q,*}(Q, N)$, which uses an ordinary free resolution of Q as a Λ-module. We will show that the two methods lead to the same action. This second method is useful for setting up the change-of-rings spectral sequence.

As an application of (4.3), we will obtain the change-of-rings spectral sequence

$$E_2^{p,q} = \mathrm{Ext}^p_{\Omega}(P, \overline{\mathrm{Ext}}^{\Gamma}_{-q}(Q, N)) \Rightarrow \mathrm{Ext}^{p+q}_{\Lambda}(P \otimes Q, N) \tag{4.4}$$

as the spectral sequence of a bisimplicial Λ-module. This will allow us in the next chapter, when we assume appropriate coproducts on P and Q and a product on N, to obtain (4.4) as the spectral sequence of a bisimplicial coalgebra. In this way we will be able to apply the results of Chapter 2, obtaining products and Steenrod operations in the spectral sequence.

Certain cases of the constructions (4.3),(4.4) are of course classical. If Q is a trivial Λ-module the work goes back to Cartan-Eilenberg [**15**]. That treatment does not assume Γ, Λ, and Ω to be Hopf algebras ... only that they are augmented algebras, that Γ is a normal subalgebra of Λ, and that Λ is free as a right Γ-module. But if one wants Steenrod operations acting on the cohomology of Λ, and on the change-of-rings spectral sequence, one must assume (4.1) an extension of Hopf algebras even if Q is trivial.

If (4.1) is an extension of group rings arising from a short-exact sequence of groups, then (4.3) and (4.4) are again classical and go back to [**29, 41**]. A more recent exposition can be found in [**43**]. That (4.3) can be generalized from the case in which Γ, Λ, Ω are group rings to the case in which they are arbitrary cocommutative Hopf algebras is to be expected. Nevertheless we feel that the details are worth writing out. To the reader who may doubt this, we issue a challenge. Where in the proof of (4.3) is cocommutativity required?

REMARK 4.2. If (4.2) is an extension, then Γ is a "right normal" subalgebra of Λ ... i.e., $I(\Gamma) \cdot \Lambda \subseteq \Lambda \cdot I(\Gamma)$, where $I(\Gamma) = \ker(\epsilon : \Gamma \to l)$ is the augmentation ideal of Γ. This follows easily from the fact that $\beta : \Lambda \to \Omega$ is a morphism of algebras. But there is another possible definition of "extension" in which Γ is left normal as a subalgebra of Λ. In this definition, one assumes a mapping $\lambda : \Lambda \to \Gamma \otimes \Omega$ that is simultaneously an isomorphism of left Γ-modules and right Ω-comodules. We note that the main results of this chapter are valid, with one very mild additional hypothesis, under this alternative definition. We say more about this in Remark 4.37 at the end of the chapter.

1. Convolutions and Conjugations

This section prepares for the study of extensions but does not consider them per se. Rather, our purpose is to define the action of a Hopf algebra on itself by conjugation, and to derive some of its properties. The main results are Propositions 4.17, 4.20 and 4.22.

We consider algebras, coalgebras and Hopf algebras over an arbitrary ground field l and graded on an abelian group $\mathbb{S}$, and modules over these. The conventions for such objects that we gave in Chapter 1 are in force here. We remind the reader also of our notation (1.2) for the permutation of the vector spaces in a tensor product. If Γ is a coalgebra with comultiplication $\psi : \Gamma \to \Gamma \otimes \Gamma$ we will write $\psi^3 : \Gamma \to \Gamma^{\otimes 3}$ for the function $\psi^3 = (\Gamma \otimes \psi)\psi = (\psi \otimes \Gamma)\psi$. If Ω is an algebra with multiplication $\mu : \Omega \otimes \Omega \to \Omega$, we will write $\mu^{\otimes 3} : \Omega^{\otimes 3} \to \Omega$ for the function $\mu^{\otimes 3} = \mu(\mu \otimes \Omega) = \mu(\Omega \otimes \mu)$. Functions $\psi^n : \Gamma \to \Gamma^{\otimes n}$ and $\mu^{\otimes n} : \Omega^{\otimes n} \to \Omega$ are defined analogously, for each $n \geq 2$. If $(\Gamma, \Delta, \ldots, \Theta)$ is a finite set of algebras we write $i_\Gamma : \Gamma \to \Gamma \otimes \Delta \otimes \cdots \otimes \Theta$ for the inclusion

$$\Gamma \simeq \Gamma \otimes l \otimes \cdots \otimes l \xrightarrow{\Gamma \otimes \eta \otimes \cdots \otimes \eta} \Gamma \otimes \Delta \otimes \cdots \otimes \Theta,$$

where each η is the unit of an algebra. Similarly $i_\Delta : \Delta \to \Gamma \otimes \Delta \otimes \cdots \otimes \Theta$, etc. If $(\Gamma, \Delta, \ldots, \Theta)$ is a finite set of coalgebras we write $p_\Gamma : \Gamma \otimes \Delta \otimes \cdots \otimes \Theta \to \Gamma$ for the projection

$$\Gamma \otimes \Delta \otimes \cdots \otimes \Theta \xrightarrow{\Gamma \otimes \epsilon \otimes \cdots \otimes \epsilon} \Gamma \otimes l \otimes \cdots \otimes l \simeq \Gamma,$$

where each ϵ is the counit of a coalgebra. Similarly $p_\Delta : \Gamma \otimes \Delta \otimes \cdots \otimes \Theta \to \Delta$, etc.

If Θ is a coalgebra and Δ an algebra, the set $\mathrm{Hom}^0(\Theta, \Delta)$ of degree-preserving l-linear maps is a semigroup under the convolution product (1.10). Lemmas 4.3 through 4.11 give some useful properties of the convolution. All are easily proved.

LEMMA 4.3. *Suppose that $\rho : \Delta \to \Delta'$ is a homomorphism of algebras, and suppose that $\zeta : \Theta' \to \Theta$ is a homomorphism of coalgebras. Then the induced mappings $\rho_* : \mathrm{Hom}^0(\Theta, \Delta) \to \mathrm{Hom}^0(\Theta, \Delta')$ and $\zeta^* : \mathrm{Hom}^0(\Theta, \Delta) \to \mathrm{Hom}^0(\Theta', \Delta)$ are mappings of semigroups. If $\tau : \Delta \to \Delta'$ is an anti-homomorphism of algebras, then the induced $\tau_* : \mathrm{Hom}^0(\Theta, \Delta) \to \mathrm{Hom}^0(\Theta, \Delta')$ is an anti-homomorphism of semigroups.*

The last statement requires the cocommutativity of Θ.

The next two lemmas are immediate corollaries of the preceding one. Recall that the symbol f^{-1} stands for the two-sided inverse of $f \in \mathrm{Hom}^0(\Theta, \Delta)$ under convolution product, if such an inverse exists. Recall also ((1.26)) that if Θ is a Hopf algebra then the two-sided inverse of the identity map in $\mathrm{Hom}^0(\Theta, \Theta)$ is called the antipode, and is written $\chi : \Theta \to \Theta$.

LEMMA 4.4. *Suppose that $f \in \mathrm{Hom}^0(\Theta, \Delta)$ is such that f^{-1} exists. Suppose that $\rho : \Delta \to \Delta'$ is a homomorphism of algebras, and that $\zeta : \Theta' \to \Theta$ is a homomorphism of coalgebras.Then both $(\rho f)^{-1}$ and $(f\zeta)^{-1}$ exist, and are given by the compositions: $(\rho f)^{-1} = \rho f^{-1}$; $(f\zeta)^{-1} = f^{-1}\zeta$.*

LEMMA 4.5. *Suppose Δ and Δ' are Hopf algebras, and that $\rho : \Delta \to \Delta'$ is a homomorphism of algebras. Then in $\mathrm{Hom}^0(\Delta, \Delta')$ the element ρ has a two-sided inverse under convolution given by $\rho^{-1} = \rho\chi$. Suppose that Θ' and Θ are Hopf algebras and that $\zeta : \Theta' \to \Theta$ is a homomorphism of coalgebras. Then in $\mathrm{Hom}^0(\Theta', \Theta)$, the element ζ has a two-sided inverse under convolution given by $\zeta^{-1} = \chi\zeta$.*

Notice that if Θ is any cocommutative Hopf algebra then $\chi : \Theta \to \Theta$ is a homomorphism of coalgebras, so we have from Lemma 4.5:

$$\chi\chi = \Theta. \tag{4.5}$$

LEMMA 4.6. *Suppose given a coalgebra Θ and an algebra Δ, and suppose given $f, g \in \mathrm{Hom}^0(\Theta, \Delta)$. Then one has:*

$$f * g = \mu(i_{\Delta_1} f * i_{\Delta_2} g). \tag{4.6}$$

Here $i_{\Delta_1} : \Delta \to \Delta \otimes \Delta$ is the inclusion of Δ as the first factor of the tensor product; and $i_{\Delta_2} : \Delta \to \Delta \otimes \Delta$ the inclusion of Δ as the second factor.

LEMMA 4.7. *Suppose given* $f \in \mathrm{Hom}^0(\Omega \otimes \Theta, \Delta)$ *and* $g \in \mathrm{Hom}^0(\Upsilon \otimes \Theta, \Delta)$. *Then in the semigroup* $\mathrm{Hom}^0(\Omega \otimes \Upsilon \otimes \Theta, \Delta)$, *the composition:*

$$\Omega \otimes \Upsilon \otimes \Theta \xrightarrow{\Omega\otimes\Upsilon\otimes\psi} \Omega \otimes \Upsilon \otimes \Theta \otimes \Theta \xrightarrow{(1,3,2,4)} \Omega \otimes \Theta \otimes \Upsilon \otimes \Theta \xrightarrow{f\otimes g} \Delta \otimes \Delta \xrightarrow{\mu} \Delta$$

can be written $f(\Omega \otimes \epsilon \otimes \Theta) * g(\epsilon \otimes \Upsilon \otimes \Theta)$.

LEMMA 4.8. *Let* Δ *be a Hopf algebra. Then in* $\mathrm{Hom}^0(\Delta^{\otimes n}, \Delta)$ *one has:*

$$\mu^n = p_{\Delta_1} * p_{\Delta_2} * \cdots * p_{\Delta_n}.$$

Here p_{Δ_k} *is the projection onto the k'th factor of the tensor product.*

LEMMA 4.9. *Let* Θ *be a coalgebra, and let* Δ, Δ' *be algebras. Suppose we are given* $f \in \mathrm{Hom}^0(\Theta, \Delta)$ *and* $g \in \mathrm{Hom}^0(\Theta, \Delta')$. *Then in* $\mathrm{Hom}^0(\Theta, \Delta \otimes \Delta')$ *one has:*

$$i_\Delta f * i_{\Delta'} g = i_{\Delta'} g * i_\Delta f.$$

LEMMA 4.10. *Let* Θ *be a coalgebra. Suppose that* Δ *is an algebra, and that* Δ' *is a central sub-algebra with inclusion map* $\alpha : \Delta' \to \Delta$. *Given arbitrary elements* $f \in \mathrm{Hom}^0(\Theta, \Delta')$ *and* $g \in \mathrm{Hom}^0(\Theta, \Delta)$ *then* $\alpha f * g = g * \alpha f$ *in* $\mathrm{Hom}^0(\Theta, \Delta)$.

LEMMA 4.11. *Let* Θ, Υ *be coalgebras; let* Ω, Δ *be algebras. Suppose we are given* $f \in \mathrm{Hom}^0(\Theta, \Omega)$ *and* $g \in \mathrm{Hom}^0(\Upsilon, \Delta)$ *for which both* f^{-1} *and* g^{-1} *exist. Then* $f \otimes g \in \mathrm{Hom}^0(\Theta \otimes \Upsilon, \Omega \otimes \Delta)$ *also has a two-sided inverse under convolution, given by* $(f \otimes g)^{-1} = f^{-1} \otimes g^{-1}$.

The proofs of all these statements are very easy; we note only that the proofs of Lemmas 4.9 and 4.10 require the (implicitly assumed) cocommutativity of Θ.

The ideas we have reviewed in this section can be combined to define the action of any Hopf algebra upon itself by conjugation. If Λ is a Hopf algebra we define $\sigma_\Lambda \in \mathrm{Hom}^0(\Lambda \otimes \Lambda, \Lambda)$ by setting:

$$\sigma_\Lambda = p_{\Lambda_2}^{-1} * p_{\Lambda_1} * p_{\Lambda_2}. \tag{4.7}$$

The notation here is like that in (4.6): p_{Λ_i} is the projection of $\Lambda \otimes \Lambda$ to the i'th copy of Λ. Note that Lemma 4.11 guarantees the existence of $(p_{\Lambda_2})^{-1}$, and gives for it the explicit formula: $(p_{\Lambda_2})^{-1} = \epsilon \otimes \chi$. Thus σ_Λ can be written as the following composition (where the notation $(2,1,3)$ is as described in (1.2)):

$$\Lambda \otimes \Lambda \xrightarrow{\Lambda\otimes\psi} \Lambda \otimes \Lambda \otimes \Lambda \xrightarrow{(2,1,3)} \Lambda \otimes \Lambda \otimes \Lambda \xrightarrow{\chi\otimes\Lambda\otimes\Lambda} \Lambda \otimes \Lambda \otimes \Lambda \xrightarrow{\mu^3} \Lambda. \tag{4.8}$$

PROPOSITION 4.12. *Let* Λ *be a Hopf algebra. The mapping* $\sigma_\Lambda : \Lambda \otimes \Lambda \to \Lambda$ *defined by* (4.7) *gives* Λ *the structure of a right module over itself.*

PROOF. The proof that $\sigma_\Lambda(\sigma_\Lambda \otimes \Lambda) = \sigma_\Lambda(\Lambda \otimes \mu)$ in $\mathrm{Hom}^0(\Lambda \otimes \Lambda \otimes \Lambda, \Lambda)$ is a straightforward diagram chase that uses (4.8), and the facts that $\mu : \Lambda \otimes \Lambda \to \Lambda$ is a morphism of coalgebras; and that $\chi : \Lambda \to \Lambda$ is an antiautomorphism of algebras. □

PROPOSITION 4.13. *Let* Λ *and* Λ' *be Hopf algebras, regarded as modules over themselves by conjugation. Regard the Hopf algebra* $\Lambda \otimes \Lambda'$ *as a module over itself by conjugation. Then* $\Lambda \otimes \Lambda'$ *is the tensor product of the* Λ*-module* Λ *with the* Λ'*-module* Λ', *in the sense of Definition 1.14.*

PROOF. This is an easy diagram chase starting from (4.8). □

The action by conjugation of a Hopf algebra upon itself is natural with respect to homomorphisms of Hopf algebras, in the following sense:

PROPOSITION 4.14. *Let $\rho : \Lambda \to \Lambda'$ be a homomorphism of Hopf algebras; and let $\sigma_\Lambda : \Lambda \otimes \Lambda \to \Lambda$ and $\sigma_{\Lambda'} : \Lambda' \otimes \Lambda' \to \Lambda'$ be the actions by conjugation of Λ and Λ' upon themselves. Then $\rho\sigma_\Lambda = \sigma_{\Lambda'}(\rho \otimes \rho)$.*

This is immediate from the definitions.

PROPOSITION 4.15. *Let Λ be a Hopf algebra, regarded as a right module over itself by conjugation. Let Λ act by conjugation on each factor of $\Lambda \otimes \Lambda$, and diagonally on the tensor product. Then the product $\mu : \Lambda \otimes \Lambda \to \Lambda$ is a homomorphism of Λ-modules.*

PROOF. In $\mathrm{Hom}^0(\Lambda \otimes \Lambda \otimes \Lambda, \Lambda)$ we must show that:

$$\sigma_\Lambda(\mu \otimes \Lambda) = \mu(\sigma_\Lambda \otimes \sigma_\Lambda)(1,3,2,4)(\Lambda \otimes \Lambda \otimes \psi). \tag{4.9}$$

But from Lemma 4.7 and (4.7) we find that the right-hand side of (4.9) is:

$$\begin{aligned} \mu(\sigma_\Lambda \otimes \sigma_\Lambda)(1,3,2,4) &= \sigma_\Lambda(\Lambda \otimes \epsilon \otimes \Lambda) * \sigma_\Lambda(\epsilon \otimes \Lambda \otimes \Lambda) \\ &= (p_{\Lambda_3}^{-1} * p_{\Lambda_1} * p_{\Lambda_3}) * (p_{\Lambda_3}^{-1} * p_{\Lambda_2} * p_{\Lambda_3}). \end{aligned} \tag{4.10}$$

On the other hand, substituting from (4.8) we find that the left-hand side of (4.9) can be written:

$$\sigma_\Lambda(\mu \otimes \Lambda) = (\epsilon \otimes \epsilon \otimes \chi) * (\mu^3) = p_{\Lambda_3}^{-1} * \mu^3,$$

so applying Lemma 4.8 we have:

$$\sigma_\Lambda(\mu \otimes \Lambda) = p_{\Lambda_3}^{-1} * (p_{\Lambda_1} * p_{\Lambda_2} * p_{\Lambda_3}). \tag{4.11}$$

But the associativity of the convolution product assures that the right hand sides of (4.10) and (4.11) are equal, so we are done. □

The above proposition can be interpreted as saying that a Hopf algebra Λ, acting on itself by conjugation, becomes a right algebra over itself, in the sense of Definition 1.26.

PROPOSITION 4.16. *Let Λ be a Hopf algebra, regarded as a right module over itself by conjugation. Let Λ act by conjugation on each factor of $\Lambda \otimes \Lambda$, and diagonally on the tensor product. Then the coproduct $\psi : \Lambda \to \Lambda \otimes \Lambda$ is a homomorphism of Λ-modules.*

Since our coproducts are commutative, ψ is a homomorphism of Hopf algebras, so this result follows from Proposition 4.14.

The above proposition can be interpreted as saying that a Hopf algebra Λ, acting on itself by conjugation, becomes a right coalgebra over itself, in the sense of Definition 1.26.

Recall the notion of a Hopf algebra over a bialgebra, reviewed in Definition 1.26.

PROPOSITION 4.17. *Let Λ be a Hopf algebra, regarded as a module over itself by conjugation. Then Λ is a Hopf algebra over itself.*

PROOF. In view of Propositions 4.12, 4.15 and 4.16, it remains only to show that the antiautomorphism $\chi : \Lambda \to \Lambda$ is a homomorphism of Λ-modules. So we must show that in $\mathrm{Hom}^0(\Lambda \otimes \Lambda, \Lambda)$ one has:

$$\chi\sigma_\Lambda = \sigma_\Lambda(\chi \otimes \Lambda). \tag{4.12}$$

But χ is an antiautomorphism of algebras so from Lemma 4.3 we have for the left hand side of (4.12): $\chi\sigma_\Lambda = \chi(p_{\Lambda_2}^{-1} * p_{\Lambda_1} * p_{\Lambda_2}) = \chi p_{\Lambda_2} * \chi p_{\Lambda_1} * \chi p_{\Lambda_2}^{-1} = p_{\Lambda_2}^{-1} * p_{\Lambda_1}^{-1} * p_{\Lambda_2}$, where we have used Lemma 4.5 in the last step. On the other hand, since Λ is cocommutative, χ and $\chi \otimes \Lambda$ are coalgebra homomorphisms, so we have from Lemma 4.3: $\sigma_\Lambda(\chi \otimes \Lambda) = (p_{\Lambda_2}^{-1} * p_{\Lambda_1} * p_{\Lambda_2})(\chi \otimes \Lambda) = p_{\Lambda_2}^{-1}(\chi \otimes \Lambda) * p_{\Lambda_1}(\chi \otimes \Lambda) * p_{\Lambda_2}(\chi \otimes \Lambda) = p_{\Lambda_2}^{-1} * p_{\Lambda_1}^{-1} * p_{\Lambda_2}$. So (4.12) is proved. □

If Θ is a Hopf algebra and if Ξ is a right Θ-algebra, we have from Definition 1.32 the notion of a $\Theta - \Xi$ module. Since we have now defined Λ as a right Λ-algebra, we can study the case $\Theta = \Xi = \Lambda$. Is there a naturally defined class of right $\Lambda - \Lambda$ modules? The next three results establish that *every* right Λ-module is a right $\Lambda - \Lambda$ module.

LEMMA 4.18. *Let Λ be a Hopf algebra, and let $\sigma_\Lambda : \Lambda \otimes \Lambda \to \Lambda$ be defined by (4.7). Then the following diagram commutes:*

$$\begin{array}{ccccc}
\Lambda \otimes \Lambda & \xrightarrow{\Lambda \otimes \psi} & \Lambda \otimes \Lambda \otimes \Lambda & \xrightarrow{(2,1,3)} & \Lambda \otimes \Lambda \otimes \Lambda \\
\downarrow{\scriptstyle \mu} & & & & \downarrow{\scriptstyle \Lambda \otimes \sigma_\Lambda} \\
\Lambda & & \xleftarrow[\mu]{\qquad\qquad} & & \Lambda \otimes \Lambda.
\end{array} \tag{4.13}$$

PROOF. The clockwise route can be written in $\mathrm{Hom}^0(\Lambda \otimes \Lambda, \Lambda)$ as $p_2 * \sigma_\Lambda$. But this is $p_2 * \sigma_\Lambda = p_2 * (p_2^{-1} * p_1 * p_2) = p_1 * p_2 = \mu$, where we have used Lemma 4.8. □

LEMMA 4.19. *Let Λ be a Hopf algebra, and let Q be any right Λ-module with structure map $\sigma_Q : Q \otimes \Lambda \to Q$. Then with $\sigma_\Lambda : \Lambda \otimes \Lambda \to \Lambda$ defined by (4.7), the following diagram commutes:*

$$\begin{array}{ccccc}
Q \otimes \Lambda \otimes \Lambda & \xrightarrow{Q \otimes \mu} & Q \otimes \Lambda & \xrightarrow{\sigma_Q} & Q \\
\downarrow{\scriptstyle Q \otimes \Lambda \otimes \psi} & & & & \\
Q \otimes \Lambda \otimes \Lambda \otimes \Lambda & & & & \uparrow{\scriptstyle \sigma_Q} \\
\downarrow{\scriptstyle (1,3,2,4)} & & & & \\
Q \otimes \Lambda \otimes \Lambda \otimes \Lambda & \xrightarrow[Q \otimes \Lambda \otimes \sigma_\Lambda]{} & Q \otimes \Lambda \otimes \Lambda & \xrightarrow[Q \otimes \mu]{} & Q \otimes \Lambda.
\end{array} \tag{4.14}$$

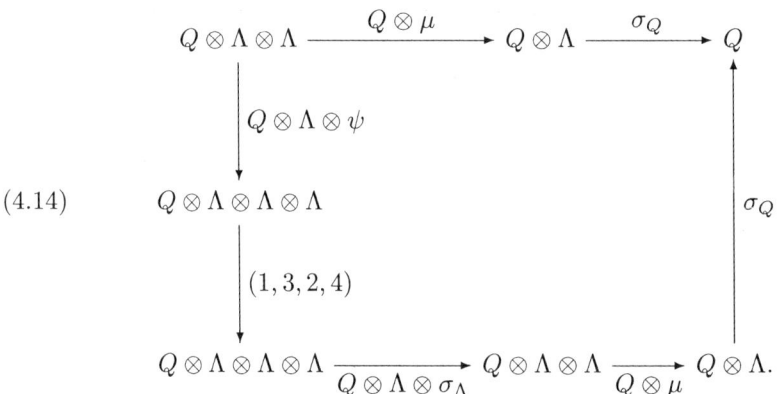

PROOF. The commutativity of (4.14) follows at once from that of (4.13). □

PROPOSITION 4.20. *Let Λ be a Hopf algebra, and let Q be an arbitrary right Λ-module. Then with Λ regarded as a right Λ-algebra under the action σ_Λ, Q becomes a right $\Lambda - \Lambda$ module.*

PROOF. We must show that the action $\sigma_Q : Q \otimes \Lambda \to Q$ is a mapping of Λ-modules, where Λ is acting on itself through σ_Λ, and then acting diagonally on $Q \otimes \Lambda$. In the diagram (4.14) we replace the top row by the composition $\sigma_Q(\sigma_Q \otimes \Lambda)$. Similarly we replace the composition $\sigma_Q(Q \otimes \mu)(Q \otimes \Lambda \otimes \sigma_\Lambda)$ by $\sigma_Q(\sigma_Q \otimes \sigma_\Lambda)$. The result is a commutative diagram that proves the proposition. □

Given a Hopf algebra Π and a right Π-algebra Ξ, we considered in Definition 3.37 the notion of a module N with "compatible" right actions of Π and Ξ. But we know that an arbitrary Hopf algebra Λ is a right algebra over itself, with conjugation action σ_Λ. So we can apply Definition 3.37 in the case $\Pi = \Xi = \Lambda$, and ask: are there right Λ-modules N for which the action of Λ is compatible with itself? We will establish in Proposition 4.22 that *every* right Λ-module is a module with self-compatible right action of Λ.

We begin by recording a useful property of σ_Λ.

LEMMA 4.21. *Let Λ be a Hopf algebra, and $\sigma_\Lambda : \Lambda \otimes \Lambda \to \Lambda$ be defined by (4.7). Then in $\mathrm{Hom}^0(\Lambda \otimes \Lambda, \Lambda)$ we have: $\sigma_\Lambda(\Lambda \otimes \chi) = p_{\Lambda_2} * p_{\Lambda_1} * p_{\Lambda_2}^{-1}$.*

PROOF. Since Λ is cocommutative, χ is an automorphism of coalgebras. Then so is $\Lambda \otimes \chi$. So from Lemma 4.3 we have:

$$\begin{aligned}\sigma_\Lambda(\Lambda \otimes \chi) &= p_{\Lambda_2}^{-1}(\Lambda \otimes \chi) * p_{\Lambda_1}(\Lambda \otimes \chi) * p_{\Lambda_2}(\Lambda \otimes \chi)\\ &= (\epsilon \otimes \chi)(\Lambda \otimes \chi) * (\Lambda \otimes \epsilon)(\Lambda \otimes \chi) * (\epsilon \otimes \Lambda)(\Lambda \otimes \chi)\\ &= (\epsilon \otimes \Lambda) * (\Lambda \otimes \epsilon) * (\epsilon \otimes \chi) = p_{\Lambda_2} * p_{\Lambda_1} * p_{\Lambda_2}^{-1},\end{aligned}$$

where we have used (4.5). □

If Λ were a group, one could define a function $\tau_\Lambda : \Lambda \times \Lambda \to \Lambda \times \Lambda$ by writing $\tau_\Lambda(\lambda, \lambda') = (\lambda'\lambda(\lambda')^{-1}, \lambda')$. This function replaces the original ordered pair by another, whose product is the same as that of the original pair with the order reversed. We get an analogous construction if Λ is an arbitrary Hopf algebra, defining $\tau_\Lambda : \Lambda \otimes \Lambda \to \Lambda \otimes \Lambda$ to be the composition:

$$\Lambda \otimes \Lambda \xrightarrow{\Lambda \otimes \psi} \Lambda \otimes \Lambda \otimes \Lambda \xrightarrow{\Lambda \otimes \chi \otimes \Lambda} \Lambda \otimes \Lambda \otimes \Lambda \xrightarrow{\sigma_\Lambda \otimes \Lambda} \Lambda \otimes \Lambda.$$

It is easy to check that this composition can be written in $\mathrm{Hom}^0(\Lambda \otimes \Lambda, \Lambda \otimes \Lambda)$ in the form $i_{\Lambda_1}\sigma_\Lambda(\Lambda \otimes \chi) * i_{\Lambda_2}p_{\Lambda_2}$, and so from Lemma 4.21 we have:

$$\tau_\Lambda = i_{\Lambda_1}\sigma_\Lambda(\Lambda \otimes \chi) * i_{\Lambda_2}p_{\Lambda_2} = i_{\Lambda_1}(p_{\Lambda_2} * p_{\Lambda_1} * p_{\Lambda_2}^{-1}) * i_{\Lambda_2}p_{\Lambda_2}.$$

It follows from Lemmas 4.6 and 4.8 that if we compose τ_Λ on the left with the multiplication on Λ we obtain:

$$\mu\tau_\Lambda = (p_{\Lambda_2} * p_{\Lambda_1} * p_{\Lambda_2}^{-1}) * p_{\Lambda_2} = p_{\Lambda_2} * p_{\Lambda_1} = \mu(1,2). \tag{4.15}$$

In this sense, application of τ_Λ produces a "pair" whose product is that of the original pair with order reversed.

The following proposition can be viewed as a companion to Proposition 4.20.

PROPOSITION 4.22. *Let Λ be a Hopf algebra, regarded as an algebra over itself under the action $\sigma_\Lambda : \Lambda \otimes \Lambda \to \Lambda$ of (4.7). Let N be an arbitrary right Λ-module. Then the right action of Λ on N is self-compatible, in the sense of Definition 3.37.*

PROOF. We must verify commutativity of (3.21) in the case $\Pi = \Xi = \Lambda$, $\sigma_N = \sigma_{N'}$, $\sigma_\Xi = \sigma_\Lambda$. But this is an immediate consequence of (4.15). □

2. Some Properties of Extensions

Given an arbitrary Hopf algebra Λ we constructed in the previous section an action of Λ on itself, by conjugation. Our first aim in this section is to show that, given an extension of Hopf algebras (4.1), then the action of Λ on itself restricts to an action of Λ on Γ, the fiber of the extension. We will observe that this action gives Γ the structure of a right Λ-Hopf algebra: see Propositions 4.23 and 4.24. We will then ask whether there is a naturally occurring class of $\Lambda - \Gamma$ modules, in the sense of Definition 1.32; and whether there is a naturally occurring class of modules with "compatible" actions of Λ and Γ, in the sense of Definition 3.37. We answer these questions in Propositions 4.26 and 4.27. Armed with these results we then use the constructions of Chapter 3, Section 3 to define a right action of $\overline{\Lambda}$ on $\mathrm{Ext}_\Gamma^{q,*}(Q, N)$ for any pair of right Λ-modules Q, N and all $q \geq 0$. As described in Chapter 3, we define this action using resolutions of Q by relative projectives in the category of $\Lambda - \Gamma$ modules.

Definition 4.1 implies that every free right Λ-module is also free over the subalgebra Γ. This fact will allow us to give a second definition of the right action of $\overline{\Lambda}$ on $\mathrm{Ext}_\Gamma^{q,*}(Q, N)$: one that uses an arbitrary free resolution of Q as a Λ-module. In Proposition 4.31 we show that the two definitions are equivalent, generalizing a well-known result from the theory of group cohomology (see for example [**43**], Chapter XI, Section 9, Exercises 3,4,5). Both definitions will prove useful. The first will allow us to cite results from Chapter 3 so that, for example, when we specialize in the next chapter to the case $l = \mathbb{Z}/2$ we will be able to write down the relation of the Λ action on $\overline{\mathrm{Ext}}^{\Gamma}_{-q,*}(Q, N)$ to the action of the Steenrod squares. The second definition is useful for setting up the change-of-rings spectral sequence later in this chapter.

We will close this section by showing that all the actions of $\overline{\Lambda}$ that we have defined pass to quotient actions of $\overline{\Omega}$.

We begin by recalling that in the last section we showed that any Hopf algebra Λ is a right module over itself, acting by conjugation $\sigma_\Lambda : \Lambda \otimes \Lambda \to \Lambda$. Now given an extension (4.1), we will show that Γ is a sub Λ-module of Λ. Thus, if $\sigma \in \mathrm{Hom}^0(\Gamma \otimes \Lambda, \Lambda)$ is defined by:

$$\sigma = \sigma_\Lambda(\alpha \otimes \Lambda) = (p_\Lambda^{-1}) * \alpha p_\Gamma * p_\Lambda, \tag{4.16}$$

then our goal is to construct the dotted arrow in the following diagram:

$$
\begin{array}{ccc}
\Lambda\otimes\Lambda & \xrightarrow{\sigma_\Lambda} & \Lambda \\
{\scriptstyle \alpha\otimes\Lambda}\uparrow & \nearrow{\scriptstyle\sigma} & \uparrow{\scriptstyle\alpha} \\
\Gamma\otimes\Lambda & \xrightarrow[\sigma_\Gamma]{} & \Gamma.
\end{array}
\tag{4.17}
$$

The construction of σ_Γ was first carried out in [**76**] for the case in which the Hopf algebras are graded-connected; and then more generally in [**24**], [**63**], and [**30**].

We begin by writing $\pi : \Omega \to \Omega/\operatorname{im}\eta$ for the projection of Ω onto its quotient by the unit. We define $\overline{\rho} : \Lambda \to \Omega/(\operatorname{im}\eta)\otimes\Lambda$ to be the composition:

$$
\Lambda \xrightarrow{\psi} \Lambda\otimes\Lambda \xrightarrow{\beta\otimes\Lambda} \Omega\otimes\Lambda \xrightarrow{\pi\otimes\Lambda} \frac{\Omega}{\operatorname{im}\eta}\otimes\Lambda.
\tag{4.18}
$$

We denote by $\rho : \Lambda \to \Omega\otimes\Lambda$ the composition of the first two maps in (4.18): this is the natural structure of Λ as a left Ω-comodule that it gets from the mapping β. Since Λ is isomorphic to $\Omega\otimes\Gamma$ as a left Ω-comodule, the following sequence of vector spaces:

$$
0 \to \Gamma \xrightarrow{\alpha} \Lambda \xrightarrow{\overline{\rho}} \frac{\Omega}{\operatorname{im}\eta}\otimes\Lambda
$$

is exact. So to show the existence of σ_Γ satisfying $\alpha\sigma_\Gamma = \sigma$ it is enough to show that $\overline{\rho}\sigma = 0$. But ρ is a homomorphism of algebras, so by Lemma 4.3 we have in $\operatorname{Hom}^0(\Gamma\otimes\Lambda, \Omega\otimes\Lambda)$:

$$
\rho\sigma = (\rho p_\Lambda)^{-1} * \rho\alpha p_\Gamma * \rho p_\Lambda.
\tag{4.19}
$$

But a short diagram chase shows that:

$$
\rho\alpha p_\Gamma = i_\Lambda \alpha p_\Gamma.
\tag{4.20}
$$

Also it is immediate from the definition of ρ that it can be written as a convolution: $\rho = i_\Omega\beta * i_\Lambda$. Then by Lemma 4.9 (we are using the cocommutativty of Λ!), we can equally well write $\rho = i_\Lambda * i_\Omega\beta$. Then since $p_\Lambda : \Gamma\otimes\Lambda \to \Lambda$ is a coalgebra morphism we have:

$$
\rho p_\Lambda = i_\Lambda p_\Lambda * i_\Omega \beta p_\Lambda.
\tag{4.21}
$$

By Lemma 4.5 each of the factors $i_\Lambda p_\Lambda$ and $i_\Omega\beta p_\Lambda$ on the right-hand side of this equation have inverses under convolution. So substituting (4.20) and (4.21) into (4.19), recalling (4.16) and using Lemma 4.9 we obtain in $\operatorname{Hom}^0(\Gamma\otimes\Lambda, \Omega\otimes\Lambda)$:

$$
\begin{aligned}
\rho\sigma &= i_\Omega(\beta p_\Lambda)^{-1} * i_\Lambda p_\Lambda^{-1} * i_\Lambda\alpha p_\Gamma * i_\Lambda p_\Lambda * i_\Omega\beta p_\Lambda \\
&= i_\Omega(\beta p_\Lambda)^{-1} * i_\Lambda\sigma * i_\Omega\beta p_\Lambda = i_\Lambda\sigma * i_\Omega(\beta p_\Lambda)^{-1} * i_\Omega\beta p_\Lambda = i_\Lambda\sigma.
\end{aligned}
\tag{4.22}
$$

(4.22) implies at once that $\overline{\rho}\sigma = 0$, so we have proved the existence of the dotted arrow in (4.17):

PROPOSITION 4.23. *Associated to the extension* (4.1) *of Hopf algebras is a unique l-linear $\sigma_\Gamma : \Gamma\otimes\Lambda \to \Gamma$ satisfying in* $\operatorname{Hom}^0(\Gamma\otimes\Lambda, \Lambda)$ *the relation:*

$$
\alpha\sigma_\Gamma = \sigma = \sigma_\Lambda(\alpha\otimes\Lambda) = p_\Lambda^{-1} * \alpha p_\Gamma * p_\Lambda.
\tag{4.23}
$$

We claim further:

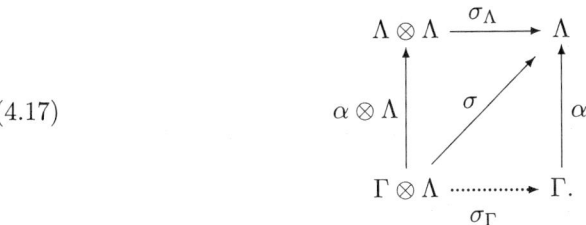

PROPOSITION 4.24. *$\sigma_\Gamma : \Gamma \otimes \Lambda \to \Gamma$ gives Γ the structure of a right Λ-Hopf algebra, in the sense of Definition 1.26.*

PROOF. Note that α is a monomorphism, and that $\alpha\sigma_\Gamma$ is the restriction to $\Gamma \otimes \Lambda$ of $\sigma_\Lambda : \Lambda \otimes \Lambda \to \Lambda$. So our result follows from Proposition 4.17. □

PROPOSITION 4.25. *Suppose Γ is central as a subalgebra of Λ. Then the action of Λ upon Γ is trivial; that is, $\sigma_\Gamma = p_\Gamma$ in $\mathrm{Hom}^0(\Gamma \otimes \Lambda, \Gamma)$.*

PROOF. From Lemma 4.10 and (4.23) we have in this case: $\alpha\sigma_\Gamma = \sigma = \alpha p_\Gamma$, so our result follows from the fact that α is monic. □

We return to the case in which (4.1) is a general extension of cocommutative Hopf algebras. We have in particular from Proposition 4.24 that Γ is a right Λ-algebra. So it is natural to ask whcther there is a naturally occurring class of right $\Lambda - \Gamma$ modules, in the sense of Definition 1.32. An answer is provided by:

PROPOSITION 4.26. *Suppose given an extension of Hopf algebras* (4.1), *and consider Γ as a right Λ-algebra under the action σ_Γ. If Q is any right Λ-module, regard it also as a right Γ-module by restriction. Then Q is a right $\Lambda - \Gamma$ module.*

PROOF. We must show that the action $Q \otimes \Gamma \to Q$ is a homomorphism of Λ-modules. But the action of Λ on $Q \otimes \Gamma$ is the restriction of the action of Λ on $Q \otimes \Lambda$ that we studied in Proposition 4.20, so our result follows from that earlier one. □

In the same spirit we have the following proposition, which generalizes Example 3.38.

PROPOSITION 4.27. *Suppose given an extension of Hopf algebras* (4.1), *and consider Γ as a right Λ-algebra under the action σ_Γ. If N is any right Λ-module, regard it also as a right Γ-module, by restriction. Then the actions of Λ and Γ upon N are compatible, in the sense of Definition 3.37.*

PROOF. This is an immediate corollary of Propositions 3.39 and 4.22. □

These last two results enable us to apply Proposition 3.42 with $\Pi = \Lambda$ and $\Xi = \Gamma$. We obtain:

PROPOSITION 4.28. *Suppose given an extension of Hopf algebras* (4.1), *and let Q, N be right Λ-modules. Regard Q and N as right Γ-modules, by restriction. Then $\mathrm{Hom}_\Gamma(Q, N)$ becomes a right module over $\overline{\Lambda}$ by the rule:*

$$(f \cdot \overline{\lambda})(q) = (-1)^{|\lambda||q|} \sum_i f\big(q(\chi\lambda_i)\big)\lambda_i' \tag{4.24}$$

for all $f \in \mathrm{Hom}_\Gamma(Q, N)$, $\lambda \in \Lambda$, $q \in Q$. This action is natural in the sense that if $g : Q' \to Q$ is a homomorphism of Λ-modules, homogeneous of degree zero, and if $h : N \to N'$ is a homomorphism of Λ-modules homogeneous of any degree, then the induced mapping $< g, h >: \mathrm{Hom}_\Gamma(Q, N) \to \mathrm{Hom}_\Gamma(Q', N')$ is a homomorphism of right $\overline{\Lambda}$-modules.

Now we have as a special case of Definition 3.45:

DEFINITION 4.29. Suppose given an extension of Hopf algebras (4.1), and let Q, N be right Λ-modules. Regard Q and N as right Γ-modules, by restriction. Then if $\mathcal{P}(\Gamma, Q) \to Q \to 0$ is any resolution of Q by relative projectives in the category of $\Lambda - \Gamma$ modules, the equation

$$\operatorname{Ext}_\Gamma^{q,*}(Q,N) = H^{q,*} \operatorname{Hom}_\Gamma(\mathcal{P}(\Gamma,Q),N) \tag{4.25}$$

defines $\operatorname{Ext}_\Gamma^{q,*}(Q,N)$ as the q'th cohomology group of a cochain complex of right $\overline{\Lambda}$-modules; and consequently defines $\operatorname{Ext}_\Gamma^{q,*}(Q,N)$ as a right $\overline{\Lambda}$-module, for each $q \geq 0$.

REMARK 4.30. From Proposition 4.24 we have that Γ is a right Λ-Hopf algebra under the conjugation action σ_Γ. Therefore Example 1.110 applies. We can form the simplicial object $W^\Gamma(Q)$ over the category of $\Lambda - \Gamma$ modules, and conclude that the associated chain complex $CW^\Gamma(Q)$ is a resolution of Q by relative projectives in the category of $\Lambda - \Gamma$ modules. In this case (4.25) becomes:

$$\operatorname{Ext}_\Gamma^{q,*}(Q,N) = H^{q,*} \operatorname{Hom}_\Gamma(CW^\Gamma(Q),N), \tag{4.26}$$

an isomorphism of right $\overline{\Lambda}$-modules; with the action of $\overline{\Lambda}$ on $\operatorname{Hom}_\Gamma(C_qW^\Gamma(Q),N)$ as in (4.24).

Proposition 4.28 and Definition 4.29 are a special case of the theory of Chapter 3, Section 3. But this special case has a feature not present in general, that provides another way of defining an action of $\overline{\Lambda}$ on $\operatorname{Ext}_\Gamma^{q,*}(Q,N)$. Proposition 4.28 gives an action of $\overline{\Lambda}$ on $\operatorname{Hom}_\Gamma(Q,N)$ for *all* right Λ-modules Q, N. So we can let $\mathcal{P}(\Lambda,Q) \to Q \to 0$ be any free resolution of Q as a right Λ-module. Then $\operatorname{Hom}_\Gamma(\mathcal{P}_q(\Lambda,Q),N)$ is a right $\overline{\Lambda}$-module for each $q \geq 0$, and by Proposition 4.28 the cochain complex $\operatorname{Hom}_\Gamma(\mathcal{P}(\Lambda,Q),N)$ is a complex of $\overline{\Lambda}$-modules. So we can define for each $q \geq 0$ a right $\overline{\Lambda}$-module $Ext_\Gamma^{q,*}(Q,N)$ by writing:

$$Ext_\Gamma^{q,*}(Q,N) = H^{q,*} \operatorname{Hom}_\Gamma(\mathcal{P}(\Lambda,Q),N). \tag{4.27}$$

Now it follows from Definition 4.1 that any free right Λ-module, when viewed as a Γ-module by restriction, is also free over Γ. Hence $\mathcal{P}(\Lambda,Q)$ can be regarded as a free resolution of Q as a right Γ-module. So we have for each $q \geq 0$ a natural isomorphism of $\mathbb{S}$-graded vector spaces:

$$\phi^q : \operatorname{Ext}_\Gamma^{q,*}(Q,N) \to Ext_\Gamma^{q,*}(Q,N). \tag{4.28}$$

But when we regard our Ext groups as right $\overline{\Lambda}$-modules, the definitions (4.25) and (4.27) look quite different:

1. The modules $\mathcal{P}_q(\Gamma,Q)$ are relative projectives in the category of $\Lambda - \Gamma$ modules, but the modules $\mathcal{P}_q(\Lambda,Q)$ not necessarily so;
2. The resolution $\mathcal{P}(\Gamma,Q) \to Q \to 0$ has a contracting homotopy that is linear over Λ, but the resolution $\mathcal{P}(\Lambda,Q) \to Q \to 0$ certainly has not;
3. The modules $\mathcal{P}_q(\Lambda,Q)$ have the property that the right action of Γ is obtained by restricting the Λ-action, but the modules $\mathcal{P}_q(\Gamma,Q)$ do not have this property.

For these reasons it may not be immediately apparent that ϕ^q of (4.28) is an isomorphism of $\overline{\Lambda}$-modules. It is so, however; and the proof is not difficult.

PROPOSITION 4.31. *ϕ^q of (4.28) is an isomorphism of $\overline{\Lambda}$-modules, for all pairs of right Λ-modules Q, N, and each $q \geq 0$.*

PROOF. By induction on q. In the case $q = 0$ we have $\mathrm{Ext}_\Gamma^{0,*}(Q,N) = Ext_\Gamma^{0,*}(Q,N) = \mathrm{Hom}_\Gamma(Q,N)$ with common definition of the right Λ-action. Suppose $q \geq 0$ an integer for which the result has been proved for all right Λ-modules Q, N. Then if Q, N be given, we choose a short-exact sequence

$$0 \to Q' \to \mathcal{F} \to Q \to 0 \tag{4.29}$$

of right Λ-modules in which $\mathcal{F}$ is free over Λ. By restriction we can view (4.29) as a short-exact sequence of Γ-modules, in which the middle term is free over Γ. It is classical that the mappings ϕ^q are compatible with connecting morphisms. So we get a commutative diagram:

$$\begin{array}{ccc} \mathrm{Ext}_\Gamma^{q,*}(Q',N) & \xrightarrow{\Delta} & \mathrm{Ext}_\Gamma^{q+1,*}(Q,N) \\ \phi^q \downarrow & & \downarrow \phi^{q+1} \\ Ext_\Gamma^{q,*}(Q',N) & \xrightarrow[\Delta]{} & Ext_\Gamma^{q+1,*}(Q,N). \end{array}$$

Here the horizontal arrows are epimorphisms of $\mathbb{S}$-graded vector spaces (actually isomorphisms if $q > 0$), the vertical arrows are isomorphisms of $\mathbb{S}$-graded vector spaces, and ϕ^q is linear over $\overline{\Lambda}$ by the inductive hypothesis. Also, the top connecting morphism Δ is linear over $\overline{\Lambda}$, by Proposition 3.46. So the $\overline{\Lambda}$- linearity of ϕ^{q+1} will be established, and our inductive proof will be complete, if we can show that the bottom homomorphism Δ is also linear over $\overline{\Lambda}$. But that is easily accomplished. Given the short-exact sequence of Λ-modules (4.29) one can build a short exact sequence of chain complexes of Λ-modules

$$0 \to \mathcal{P}(\Lambda, Q') \to \mathcal{P}(\Lambda, \mathcal{F}) \to \mathcal{P}(\Lambda, Q) \to 0, \tag{4.30}$$

where these are free resolutions of Q', $\mathcal{F}$, and Q, respectively. Using Proposition 4.26 we can also consider (4.30) as a short-exact sequence of chain complexes of $\Lambda - \Gamma$ modules, where the action of Γ on each chain complex is obtained by restricting the action of Λ. Similarly, using Proposition 4.27 we regard N as a module with compatible right actions of Λ and Γ, where the action of Γ is obtained by restriction. Then it follows from Proposition 3.43 that $\Delta : Ext_\Gamma^{q,*}(Q',N) \to Ext_\Gamma^{q+1,*}(Q,N)$ is linear over $\overline{\Lambda}$, so our proof is complete. □

A version of Proposition 4.31 for group cohomology is given by Mac Lane [**43**], Chapter 11, Section 9, Exercises 3-5. Our result is a generalization of that one.

We are now entitled to drop the notational distinction between $\mathrm{Ext}_\Gamma^{q,*}(Q,N)$ and $Ext_\Gamma^{q,*}(Q,N)$, and will use only the former. We will have uses for each of our equivalent definitions of the action of $\overline{\Lambda}$ on $\mathrm{Ext}_\Gamma^{q,*}(Q,N)$. The original definition (4.25) comes from Chapter 3, and we will use it when we need to cite the results of that chapter...for example, to establish the relaltionship between the actions of $\overline{\Lambda}$ and of the Steenrod squares on $\mathrm{Ext}_\Gamma^{*,*}(Q,N)$. We will also use it in proving Proposition 4.34. We will need the alternative definition (4.27), and its "negative" (4.36), in setting up the change-of-rings spectral sequence in (4.47) below.

Now we pass from actions of $\overline{\Lambda}$ to actions of $\overline{\Omega}$. We begin by noting one consequence of Definition 4.1: Ω is a quotient Hopf algebra of Λ, in a sense that we now make precise. Write $I(\Gamma) = \ker(\epsilon : \Gamma \to l)$ for the kernel of the augmentation. Recall from [**51**], Definition 3.3 that an inclusion $\alpha : \Gamma \to \Lambda$ of augmented algebras is called "right normal" if $I(\Gamma) \cdot \Lambda \subseteq \Lambda \cdot I(\Gamma)$. If $\alpha : \Gamma \to \Lambda$ is a right normal inclusion of augmented algebras then the quotient vector space $\Lambda \otimes_\Gamma l$ inherits from Λ the structure of an augmented algebra. If in addition Γ and Λ are Hopf algebras then the coproduct on Λ induces a coproduct on the quotient space , under which $\Lambda \otimes_\Gamma l$ becomes a Hopf algebra. From Definition 4.1 we have easily:

LEMMA 4.32. *Suppose* (4.1) *an extension of Hopf algebras, in the sense of Definition 4.1. Then* $\alpha : \Gamma \to \Lambda$ *is a right normal inclusion of algebras; and* $\beta : \Lambda \to \Omega$ *induces an isomorphism of Hopf algebras:*

$$\Lambda \otimes_\Gamma l \cong \Omega. \tag{4.31}$$

The proof is easy, and we do not write it out.

PROPOSITION 4.33. *Suppose* (4.1) *is an extension of Hopf algebras, and suppose* Q, N *are right* Λ*-modules. Let the right actions of* $\overline{\Lambda}$ *on the modules* $\mathrm{Hom}_\Gamma(Q,N)$ *and* $\mathrm{Ext}_\Gamma^{q,*}(Q,N)$ *be defined as in* (4.24) *and* (4.25). *Then the actions of the sub-algebra* $\overline{\Gamma} \subseteq \overline{\Lambda}$ *on these modules is trivial. Therefore the actions of* $\overline{\Lambda}$ *pass to actions of the quotient* $\overline{\Omega} = \overline{\Lambda} \otimes_{\overline{\Gamma}} l$*:*

$$\mathrm{Hom}_\Gamma(Q,N) \otimes \overline{\Omega} \to \mathrm{Hom}_\Gamma(Q,N) \tag{4.32}$$

$$\mathrm{Ext}_\Gamma^{q,*}(Q,N) \otimes \overline{\Omega} \to \mathrm{Ext}_\Gamma^{q,*}(Q,N) \tag{4.33}$$

for all $q \geq 0$.

PROOF. The triviality of the action of $\overline{\Gamma}$ on $\mathrm{Hom}_\Gamma(Q,N)$ follows immediately from (4.24) and the defining property (1.26) of the antipode χ. □

We conclude this section with a remark about central extensions.

PROPOSITION 4.34. *Suppose that* Γ *is central as a subalgebra of* Λ*, and that* Q *and* N *are trivial as* Λ*-modules. Then* $\mathrm{Ext}_\Gamma^{q,*}(Q,N)$ *is trivial as a right* $\overline{\Lambda}$*-module, and as a right* $\overline{\Omega}$*-module, for all* $q \geq 0$.

PROOF. We can use (4.26) to calculate the action of $\overline{\Lambda}$ on $\mathrm{Ext}_\Gamma^{q,*}(Q,N)$. But $W_q^\Gamma(Q) = Q \otimes \Gamma^{\otimes(q+1)}$ with diagonal action of Λ. By hypothesis Λ acts trivially on Q; and by Proposition 4.25, Λ acts trivially on Γ as well. So the action of Λ on $W_q^\Gamma(Q)$ is trivial. Since we are also assuming that Λ acts trivially on N, equation (4.24) shows $\overline{\Lambda}$ acting trivially on $\mathrm{Hom}_\Gamma(W_q^\Gamma(Q), N)$ for each q. The result follows. □

3. Adjunction Isomorphism and Change-of-Rings Spectral Sequence

In this section we fix an extension of Hopf algebras (4.1); and we fix a right Ω-module P, and right Λ-modules Q and N. Our goal is to establish the adjunction isomorphism (4.3), and to derive the change-of-rings spectral sequence (4.4) as the spectral sequence of a bisimplicial Λ-module. Our arguments of course follow the classical case [**41, 29, 42**] in which the objects in (4.1) are groups. But our objects are cocommutative Hopf algebras, and we have to use the results of the previous section.

We need a preliminary remark. For the sake of consistency with the theory in Chapter 3, which is of a piece, we have so far spoken of actions of $\overline{\Lambda}$ and $\overline{\Omega}$

on $\mathrm{Hom}_\Gamma(Q, N)$. But the actions we will need for our adjunction isomorphism are obtained from these by replacing each vector space by its negative. Thus, for all right Λ-modules Q, N we define the right action of Λ on $\overline{\mathrm{Hom}}^\Gamma(Q, N)$ by replacing (4.24) by its negative:

$$(f \cdot \lambda)(q) = (-1)^{|\lambda||q|} \sum_i f\big(q(\chi\lambda_i)\big)\lambda_i' \tag{4.34}$$

for all $f \in \overline{\mathrm{Hom}}^\Gamma(Q, N)$. We need to define for each $q \geq 0$ a corresponding action of Λ on $\overline{\mathrm{Ext}}^\Gamma_{-q,*}(Q, N)$, where the latter is defined as an $\mathbb{S}$-graded vector space by (1.47). So we represent $\overline{\mathrm{Ext}}^\Gamma_{-q,*}(Q, N)$ as the $-q$'th homology group of a chain complex of Λ-modules:

$$\overline{\mathrm{Ext}}^\Gamma_{-q,*}(Q, N) = H_{-q,*}\overline{\mathrm{Hom}}^\Gamma(\mathcal{P}(\Gamma, Q), N). \tag{4.35}$$

Here $\mathcal{P}(\Gamma, Q)$ is any resolution of Q by relative projectives in the category of $\Lambda - \Gamma$ modules, and the action of Λ on $\overline{\mathrm{Hom}}^\Gamma(\mathcal{P}_q(\Gamma, Q), N)$ is as in (4.34). Equivalently we can write:

$$\overline{\mathrm{Ext}}^\Gamma_{-q,*}(Q, N) = H_{-q,*}\overline{\mathrm{Hom}}^\Gamma(\mathcal{P}(\Lambda, Q), N) \tag{4.36}$$

where $\mathcal{P}(\Lambda, Q)$ is any free resolution of Q as a right Λ-module, and the action of Λ on $\overline{\mathrm{Hom}}^\Gamma(\mathcal{P}_q(\Lambda, Q), N)$ is again given by (4.34). Formulas (4.35) and (4.36) are of course just the negatives of (4.25) and (4.27), and Proposition 4.31 says that they define the same Λ-actions on $\overline{\mathrm{Ext}}^\Gamma_{-q,*}(Q, N)$.

As in (4.32),(4.33) we observe that the actions of Γ on $\overline{\mathrm{Hom}}^\Gamma(Q, N)$ and on $\overline{\mathrm{Ext}}^\Gamma_{-q,*}(Q, N)$ are trivial, so that the original actions of Λ pass to quotients:

$$\overline{\mathrm{Hom}}^\Gamma(Q, N) \otimes \Omega \to \overline{\mathrm{Hom}}^\Gamma(Q, N) \tag{4.37}$$

$$\overline{\mathrm{Ext}}^\Gamma_{-q,*}(Q, N) \otimes \Omega \to \overline{\mathrm{Ext}}^\Gamma_{-q,*}(Q, N). \tag{4.38}$$

These are the vector space negatives of (4.32) and (4.33).

PROPOSITION 4.35. *Suppose given an extension of Hopf algebras (4.1), and suppose given a right Ω-module P, and right Λ-modules Q and N. Let Λ act on P through $\beta : \Lambda \to \Omega$, and let Λ act on $P \otimes Q$ diagonally. Then the adjunction mapping ϕ of (1.5) satisfies:*

$$\phi(\mathrm{Hom}_\Lambda(P \otimes Q, N)) \subseteq \mathrm{Hom}_\Omega(P, \overline{\mathrm{Hom}}^\Gamma(Q, N))$$

and its inverse ζ of (1.6) satisfies:

$$\zeta(\mathrm{Hom}_\Omega(P, \overline{\mathrm{Hom}}^\Gamma(Q, N))) \subseteq \mathrm{Hom}_\Lambda(P \otimes Q, N).$$

Consequently ϕ and ζ define a pair of isomorphisms of $\mathbb{S}$-graded vector spaces inverse to each other:

$$\mathrm{Hom}_\Lambda(P \otimes Q, N) \underset{\zeta}{\overset{\phi}{\rightleftarrows}} \mathrm{Hom}_\Omega(P, \overline{\mathrm{Hom}}^\Gamma(Q, N)). \tag{4.39}$$

The mappings ϕ, ζ are natural with respect to homomorphisms of degree zero of the Ω-module P and of the Λ-modules Q and N, in the obvious sense.

PROOF. We have defined the Λ-action on P in such a way that the action of Γ is trivial. So if $f \in \mathrm{Hom}_\Lambda(P \otimes Q, N)$ then for each $\gamma \in \Gamma$ we have $[(\phi f)(p)](q\gamma) = f(p \otimes q\gamma) = f((p \otimes q)\gamma) = f(p \otimes q) \cdot \gamma = [((\phi f)(p))(q)]\gamma$. Therefore $(\phi f)(p)$ is a Γ-linear homomorphism for each $p \in P$. We next check that ϕf is Ω-linear as a homomorphism from P to $\overline{\mathrm{Hom}_\Gamma}(Q, N)$. By (4.34), this is equivalent to showing:

$$f(p\lambda \otimes q) = (-1)^{|\lambda||q|} \sum_i [f(p \otimes q(\chi\lambda_i))]\lambda_i' \tag{4.40}$$

for all $f \in \mathrm{Hom}_\Lambda(P \otimes Q, N)$, $p \in P$, $q \in Q$, $\lambda \in \Lambda$. Thus, it suffices to show that:

$$p\lambda \otimes q = (-1)^{|\lambda||q|} \sum_i (p \otimes q(\chi\lambda_i))\lambda_i'. \tag{4.41}$$

But (4.41) is an easy consequence of the definition of the diagonal action on $P \otimes Q$, the commutativity and associativity of the coproduct on Λ, and the defining property (1.26) of the antipode. Finally we must show that if $g \in \mathrm{Hom}_\Omega(P, \overline{\mathrm{Hom}_\Gamma}(Q, N))$ then ζg as given by (1.6) is a homomorphism of Λ-modules. Using (4.34) and the cocommutativity of Λ we find that for each $p \in P$, $q \in Q$, $\lambda \in \Lambda$:

$$(\zeta g)[(p \otimes q)\lambda] = \sum_i [g(p)(q\lambda_i \cdot \chi\lambda_i')]\lambda_i''. \tag{4.42}$$

By (1.26), the right hand side of (4.42) reduces to $[g(p)(q)]\lambda = [(\zeta g)(p \otimes q)]\lambda$, so we are done. □

As an immediate application of this result we can set up the change-of-rings spectral sequence as the spectral sequence of a bisimplicial Λ-module.

Again we fix (4.1), an extension of Hopf algebras. Suppose we are given a right Ω-module P and a right Λ-module Q. Then we have described in Chapter 1, Section 11 simplicial objects $W^\Lambda(Q)$ and $W^\Omega(P)$ over the categories of Λ-modules and Ω-modules, respectively. We have observed that the associated chain complexes $CW^\Lambda(Q)$ and $CW^\Omega(P)$ are free resolutions of Q over Λ and of P over Ω, respectively. We can regard $W^\Omega(P)$ also as a simplicial Λ-module, with Λ acting through $\beta : \Lambda \to \Omega$. For all $p \geq 0, q \geq 0$ set

$$X_{p,q} = W_p^\Omega(P) \otimes W_q^\Lambda(Q), \tag{4.43}$$

with diagonal right action of Λ. We make the collection $\{X_{p,q} \mid p \geq 0, q \geq 0\}$ into a bisimplicial Λ-module X. The "horizontal" face and degeneracy operators are obtained from the face and degeneracy operators on $W^\Omega(P)$ by tensoring them with the identity map on $W_q^\Lambda(Q)$; similarly, the vertical simplicial operators are obtained from the simplicial operators on $W^\Lambda(Q)$. As in Chapter 1, Section 15 we can consider the associated total chain complex of Λ-modules CX. As a chain complex over the ground field l we have $CX = CW^\Omega(P) \otimes CW^\Lambda(Q)$. From the Künneth theorem and Proposition 1.24 one sees that CX is a free resolution of $P \otimes Q$ as a Λ-module. So for any right Λ-module N we have

$$H^{*,*} \mathrm{Hom}_\Lambda(CX, N) = \mathrm{Ext}_\Lambda^{*,*}(P \otimes Q, N), \tag{4.44}$$

an isomorphism of $\mathbb{Z} \times \mathbb{S}$-graded vector spaces. On the left-hand side the grading on $\mathbb{S}$, "internal degree", which was suppressed in Chapters 1 and 2, is being made explicit.

The "spectral sequence of a bisimplicial Λ-module X with coefficients in a Λ-module N" has been defined in Chapter 1, Section 16.

DEFINITION 4.36. Suppose given an extension of Hopf algebras (4.1), a right Ω-module P, and right Λ-modules Q and N. By the associated change-of-rings spectral sequence we mean the spectral sequence of the bisimplicial Λ-module X with coefficients in N, where X is as in (4.43).

By (4.44), our spectral sequence converges to $\mathrm{Ext}_{\Lambda}^{*,*}(P \otimes Q, N)$. We use (1.149) to describe the E_2-term, and for the rest of this paragraph will suppress internal degree. We begin by studying for each fixed $p \geq 0$ the cochain complex $\mathrm{Hom}_{\Lambda}(C_pW^{\Omega}(P) \otimes CW^{\Lambda}(Q), N)$, with differential induced by that on $CW^{\Lambda}(Q)$. Equation (1.45) gives an isomorphism of cochain complexes:

(4.45)
$$\phi_p : \mathrm{Hom}(C_pW^{\Omega}(P) \otimes CW^{\Lambda}(Q), N) \to \mathrm{Hom}(C_pW^{\Omega}(P), \overline{\mathrm{Hom}}(CW^{\Lambda}(Q), N)).$$

By Proposition 4.35, if ϕ_p is restricted to $\mathrm{Hom}_{\Lambda}(C_pW^{\Omega}(P) \otimes C_qW^{\Lambda}(Q), N)$ then it maps isomorphically onto $\mathrm{Hom}_{\Omega}(C_pW^{\Omega}(P), \overline{\mathrm{Hom}}^{\Gamma}(C_qW^{\Lambda}(Q), N))$. So for each $p \geq 0$, (4.45) restricts to an isomorphism of cochain complexes:

(4.46)
$$\phi_p : \mathrm{Hom}_{\Lambda}(C_pW^{\Omega}(P) \otimes CW^{\Lambda}(Q), N) \to \mathrm{Hom}_{\Omega}(C_pW^{\Omega}(P), \overline{\mathrm{Hom}}^{\Gamma}(CW^{\Lambda}(Q), N)).$$

Continuing to suppress internal degree we write $\overline{\mathrm{Ext}}^{\Gamma}_{-q}(Q, N)$ for the $\mathbb{S}$-graded vector space $\overline{\mathrm{Ext}}^{\Gamma}_{-q,*}(Q, N)$, and use (4.36) for the isomorphism of Ω-modules: $\overline{\mathrm{Ext}}^{\Gamma}_{-q}(Q, N) = H_{-q}\overline{\mathrm{Hom}}^{\Gamma}(CW^{\Lambda}(Q), N)$. Here of course the Definition 4.1 plays a crucial role: it assures that a free right Λ-module is also free over Γ. Recalling that $C_pW^{\Omega}(P)$ is free as a right Ω-module, we have from (4.46):

(4.47)
$$\begin{aligned} &H^q \mathrm{Hom}_{\Lambda}(C_pW^{\Omega}(P) \otimes CW^{\Lambda}(Q), N) \\ &\qquad = H^q \mathrm{Hom}_{\Omega}(C_pW^{\Omega}(P), \overline{\mathrm{Hom}}^{\Gamma}(CW^{\Lambda}(Q), N)) \\ &\qquad = \mathrm{Hom}_{\Omega}(C_pW^{\Omega}(P), H_{-q}\overline{\mathrm{Hom}}^{\Gamma}(CW^{\Lambda}(Q), N)) \\ &\qquad = \mathrm{Hom}_{\Omega}(C_pW^{\Omega}(P), \overline{\mathrm{Ext}}^{\Gamma}_{-q}(Q, N)). \end{aligned}$$

Now (1.149) gives a description of the E_2-term of our spectral sequence as:

(4.48)
$$\begin{aligned} E_2^{p,q} &= H^p \mathrm{Hom}_{\Omega}(CW^{\Omega}(P), \overline{\mathrm{Ext}}^{\Gamma}_{-q}(Q, N)) \\ &= \mathrm{Ext}^p_{\Omega}(P, \overline{\mathrm{Ext}}^{\Gamma}_{-q}(Q, N)). \end{aligned}$$

So we have the change-of-rings spectral sequence

(4.49)
$$E_2^{p,q} = \mathrm{Ext}^p_{\Omega}(P, \overline{\mathrm{Ext}}^{\Gamma}_{-q}(Q, N)) \Longrightarrow \mathrm{Ext}^{p+q}_{\Lambda}(P \otimes Q, N),$$

for any extension (4.1) of cocommutative Hopf algebras, any right Ω-module P, and any right Λ-modules Q, N.

It is easy to write a version of (4.49) that remembers internal degree. Since P is $\mathbb{S}$-graded, and $\overline{\mathrm{Ext}}^{\Gamma}_{-q}(Q,N)$ is $\mathbb{S}$-graded for each $q \geq 0$, so $\mathrm{Ext}^p_\Omega(P, \overline{\mathrm{Ext}}^{\Gamma}_{-q}(Q,N))$ is also $\mathbb{S}$-graded for all $p, q \geq 0$. Then (4.49) reads:

$$E_2^{p,q,*} = \mathrm{Ext}^{p,*}_\Omega(P, \overline{\mathrm{Ext}}^{\Gamma}_{-q}(Q, N)) \Longrightarrow \mathrm{Ext}^{p+q,*}_\Lambda(P \otimes Q, N).$$

REMARK 4.37. We pointed out in Remark 4.2 that one can define an "extension of Hopf algebras" in a slightly different way from the one we have chosen; postulating in place of the diagram (4.2), an analogous diagram with a mapping $\lambda : \Lambda \to \Gamma \otimes \Omega$, that is simultaneously an isomorphism of left Γ-modules and right Ω-comodules. In this case Γ is a left normal, rather than a right normal, subalgebra of Λ. That is, one has $\Lambda \cdot I(\Gamma) \subseteq I(\Gamma) \cdot \Lambda$, where $I(\Gamma) = \ker(\epsilon : \Gamma \to l)$ is the augmentation ideal of Γ. All the results of this chapter are valid "verbatim" under this alternative definition of extension, if we make only the additional assumption that Λ is free as a right Γ-module. (That Λ is free as a left Γ-module is apparent in the context of our modified definition of extension; that it is free as a right Γ-module must now be independently assumed.) The key point is that the right action of Λ upon itself by conjugation should restrict to a right action of Λ upon Γ. But the proof of Proposition 4.23 goes through without essential change. One simply replaces $\Omega \otimes \Lambda$ by $\Lambda \otimes \Omega$ in the appropriate places. The action of Λ on Γ is still from the right. Proposition 4.28, and Definition 4.29 hold, exactly as written. But to get the alternative definition of the right action of $\overline{\Lambda}$ upon $\mathrm{Ext}^{q,*}_\Gamma(Q, N)$, as in (4.27), one needs our extra assumption that Λ is free as a right module over Γ. The passage from right actions of $\overline{\Lambda}$ to right actions of $\overline{\Omega}$, as in Proposition (4.33), goes through with our modified definition of extension; one need only replace (4.31) by the statement $l \otimes_\Gamma \Lambda \cong \Omega$. The proof of the main theorem of this section, Proposition 4.35, as well as the construction of the change-of-rings spectral sequence, go through exactly as written. But the description (4.48) of the E_2-term of the spectral sequence, with our modified definition of extension, requires the extra assumption that Λ is free as a right Γ-module, since it uses (4.27) and (4.36).

CHAPTER 5

Steenrod Operations in the Change-of-Rings Spectral Sequence

We continue to study the change-of-rings spectral sequence

$$E_2^{p,q} = \operatorname{Ext}_\Omega^p(P, \overline{\operatorname{Ext}}^\Gamma_{-q}(Q, N)) \Rightarrow \operatorname{Ext}_\Lambda^{p+q}(P \otimes Q, N) \tag{5.1}$$

as set up in the previous chapter; but with added assumptions, so we can apply the theory of Chapter 2. Our extension of $\mathbb{S}$-graded Hopf algebras

$$\Gamma \xrightarrow{\alpha} \Lambda \xrightarrow{\beta} \Omega \tag{5.2}$$

is now over the ground field $l = \mathbb{Z}/2$. The right Ω-module P we now assume to be an Ω-coalgebra; the right Λ-module Q we assume is a Λ-coalgebra, and the right Λ-module N is a commutative Λ-algebra. These assumptions will be in force throughout this chapter.

We will put Steenrod operations into the spectral sequence (5.1), and compute the operations at the E_2-level. In the case in which (5.2) is a central extension and $P = Q = N = \mathbb{Z}/2$, our results reduce to those of Uehara [**86**]. The main results of this chapter are Theorem 5.2, which tells how the squaring operations act on each page of the spectral sequence and how they commute with the differentials; Proposition 5.26, which computes the diagonal squaring operations on the E_2-term of the spectral sequence; and Proposition 5.27, which computes the vertical squaring operations. In the last four sections we discuss examples, involving particular extensions. Some are completely worked out, and some are presented as open problems. In particular in Section 6 we discuss the application of our theory that has already been made by Palmieri, in his description of the cohomology of the Steenrod algebra "up to nilpotence".

1. The Spectral Sequence with its Products and Steenrod Squares

The main result of this section is Theorem 5.2, which tells how the squaring operations act on each page of the change-of-rings spectral sequence and how they commute with the differentials. This theorem also gives the compatibility of the squaring operations on E_∞ with those defined on the target of the spectral sequence.

As in Chapter 1 Section 11, the construction $W^\Omega(P)$ can now be regarded as a simplicial object over the category of Ω-coalgebras, and $W^\Lambda(Q)$ can be regarded as a simplicial object over the category of Λ-coalgebras. For all $p, q \geq 0$ we define the Λ-coalgebra $X_{p,q}$ by writing

$$X_{p,q} = W_p^\Omega(P) \otimes W_q^\Lambda(Q) \tag{5.3}$$

where Λ acts on $W_p^\Omega(P)$ through $\beta : \Lambda \to \Omega$, and the tensor product is the tensor product of Λ-coalgebras, as in Proposition 1.29. The collection $\{X_{p,q} \,|\, p, q \geq 0\}$ is a bisimplicial object X over the category of Λ-coalgebras. As in (4.44) we have an isomorphism of $\mathbb{Z} \times \mathbb{S}$-graded vector spaces

$$H^{*,*} \operatorname{Hom}_\Lambda(CX, N) = \operatorname{Ext}_\Lambda^{*,*}(P \otimes Q, N). \tag{5.4}$$

(On the left-hand side the grading on $\mathbb{S}$, "internal degree", which was suppressed in Chapters 1 and 2, is being made explicit.) But in the present context both sides of (5.4) have additional structure. Equations (2.12) and (2.17) define products and Steenrod operations

$$\mu : H^{*,*} \operatorname{Hom}_\Lambda(CX, N) \otimes H^{*,*} \operatorname{Hom}_\Lambda(CX, N) \to H^{*,*} \operatorname{Hom}_\Lambda(CX, N); \tag{5.5}$$

$$Sq^k : H^{n,s} \operatorname{Hom}_\Lambda(CX, N) \to H^{n+k,2s} \operatorname{Hom}_\Lambda(CX, N). \tag{5.6}$$

On the other hand we can apply the purely Hopf-theoretic constructions of Chapter 1, Section 14, to the case in which $\Theta = \Lambda$, and the Λ-coalgebra M is the tensor product of the Λ-coalgebras P and Q. We get a product

$$\mu : \operatorname{Ext}_\Lambda^{*,*}(P \otimes Q, N) \otimes \operatorname{Ext}_\Lambda^{*,*}(P \otimes Q, N) \to \operatorname{Ext}_\Lambda^{*,*}(P \otimes Q, N) \tag{5.7}$$

as in (1.125); and Steenrod operations

$$Sq^k : \operatorname{Ext}_\Lambda^{n,s}(P \otimes Q, N) \to \operatorname{Ext}_\Lambda^{n+k,2s}(P \otimes Q, N) \tag{5.8}$$

as in (1.126). In the remainder of this chapter we will apply the theory of Chapter 2 to discuss products and Steenrod operations in the change-of-rings spectral sequence. These operations "converge to" the products and Steenrod operations (5.5),(5.6). We will want to know that these are the same as the products and Steenrod operations (5.7), (5.8) normally defined on the cohomology of the Hopf algebra Λ. The following proposition establishes this agreement.

PROPOSITION 5.1. *Under the identification* (5.4)*, the products* (5.5) *and* (5.7) *agree; and the Steenrod squares* (5.6) *and* (5.8) *agree.*

PROOF. The operations (5.5) and (5.6) were defined by means of the mappings $K_k : C(X \times X) \to CX \otimes CX$ of (2.6). These satisfy (2.7). But X is a (commutative) bisimplicial coalgebra, so the coproduct $\psi : X \to X \times X$ induces a chain mapping $\psi : CX \to C(X \times X)$ satisfying $T\psi = \psi$. So if we set $\mathcal{D}_k = K_k \psi$ for each $k \geq 0$ we obtain a sequence of Λ-linear mappings $\mathcal{D}_k : CX \to CX \otimes CX$ satisfying all the conditions of Definition 1.121 for a "homological cup-k product". We are applying the constructions in the latter part of Section 14 of Chapter 1 with CX in the role of the free resolution $\mathcal{P}(P \otimes Q)$ of the Λ-coalgebra $P \otimes Q$. These observations imply the result. □

Now we consider the spectral sequence (5.1) of the bisimplicial Λ-coalgebra X. The theory of Chapter 2 applies at once, giving:

THEOREM 5.2. *Let* (5.2) *be a given extension of Hopf algebras over* $\mathbb{Z}/2$*; suppose given an* Ω*-coalgebra* P*, a* Λ*-coalgebra* Q*, and a commutative* Λ*-algebra* N*. Then the change-of-rings spectral sequence* (5.1) *admits products*

$$\mu : E_r^{*,*} \otimes E_r^{*,*} \to E_r^{*,*}$$

and squaring operations

$$\begin{aligned} Sq^k &: E_r^{p,q} \to E_r^{p,q+k} && \text{if } 0 \le k \le q \\ Sq^k &: E_r^{p,q} \to E_{r+k-q}^{p+k-q,2q} && \text{if } q \le k \le q+r-2 \\ Sq^k &: E_r^{p,q} \to E_{2r-2}^{p+k-q,2q} && \text{if } q+r-2 \le k \end{aligned}$$

for all $r \geq 2$, satisfying all propositions and theorems of Chapter 2, Sections 2 and 3. In particular, the operations on E_r are determined by the operations at the E_2-level as described in Theorem 2.16. Products and squaring operations commute with the differentials of the spectral sequence as described in Proposition 2.10 and Theorem 2.17. Products and squaring operations are defined on E_∞ and are compatible with the operations on the target of the spectral sequence, (5.7) *and* (5.8) *in the manner described by Proposition 2.11 and Theorem 2.22. If $\alpha \in E_r$ then α^2 survives to E_{2r-1}, as described in Theorem 2.21.*

2. Steenrod Operations on $\mathrm{Ext}_{\Omega}^{*,*}(P, \overline{\mathrm{Ext}}_*^{\Gamma}(Q, N))$

In this section we will define in purely Hopf-theoretic language a product

(5.9)

$$\mu : \mathrm{Ext}_{\Omega}^{*,*}(P, \overline{\mathrm{Ext}}_*^{\Gamma}(Q, N)) \otimes \mathrm{Ext}_{\Omega}^{*,*}(P, \overline{\mathrm{Ext}}_*^{\Gamma}(Q, N)) \to \mathrm{Ext}_{\Omega}^{*,*}(P, \overline{\mathrm{Ext}}_*^{\Gamma}(Q, N))$$

and squaring operations

(5.10a) $$Sq_V^k : \mathrm{Ext}_{\Omega}^{p,s}(P, \overline{\mathrm{Ext}}_{-q}^{\Gamma}(Q, N)) \to \mathrm{Ext}_{\Omega}^{p,2s}(P, \overline{\mathrm{Ext}}_{-q-k}^{\Gamma}(Q, N))$$

(5.10b) $$Sq_D^k : \mathrm{Ext}_{\Omega}^{p,s}(P, \overline{\mathrm{Ext}}_{-q}^{\Gamma}(Q, N)) \to \mathrm{Ext}_{\Omega}^{p+k,2s}(P, \overline{\mathrm{Ext}}_{-2q}^{\Gamma}(Q, N))$$

for all $p, q, k \geq 0$ and $s \in \mathbb{S}$. Here our gradings are those described at the end of Chapter 4. The product and diagonal operations are given in Definition 5.7, and the vertical operations in Definition 5.19. Then in Section 4 we will identify these operations with the product and with the diagonal and vertical Steenrod squares on the E_2-term of the change-of-rings spectral sequence, as these were defined in Chapter 2.

We begin by defining operations on $\mathrm{Ext}_{\Gamma}^*(Q, N)$. We apply the constructions of Chapter 1, Section 14 in the case $\Theta = \Gamma, M = Q$. To be consistent with the notation in that section we remember internal degree. We obtain a product

(5.11) $$\mu : \mathrm{Ext}_{\Gamma}^{*,*}(Q, N) \otimes \mathrm{Ext}_{\Gamma}^{*,*}(Q, N) \to \mathrm{Ext}_{\Gamma}^{*,*}(Q, N)$$

and squaring operations

(5.12) $$Sq^k : \mathrm{Ext}_{\Gamma}^{q,s}(Q, N) \to \mathrm{Ext}_{\Gamma}^{q+k,2s}(Q, N)$$

for all $k \in \mathbb{Z}$, $s \in \mathbb{S}$. As in Proposition 1.120 we observe these operations give $\mathrm{Ext}_{\Gamma}^{*,*}(Q, N)$ the structure of a commutative left algebra over the algebra with coproducts $\mathcal{H}$. The elements of $\mathrm{Ext}_{\Gamma}^{q,s}(Q, N)$ have additive degree q and multiplicative degree s in the sense of Definition 1.50.

Now we ask how the above operations are related to the right action of $\overline{\Lambda}$ on the modules $\mathrm{Ext}_{\Gamma}^{q,*}(Q, N)$, as given by Definition 4.29. This definition is a special case of the theory in Chapter 3, Section 3, with $\Pi = \Lambda$, $\Xi = \Gamma$, and $M = Q$. In fact, all the hypotheses of Proposition 3.47 are satisfied, since we have from Proposition 4.24 that the right action of Λ on Γ by conjugation makes Γ into a Λ-Hopf algebra. So we apply Proposition 3.47 to conclude:

PROPOSITION 5.3. *With the product* (5.11), $\mathrm{Ext}_\Gamma^{*,*}(Q,N)$ *becomes a commutative right* $\overline{\Lambda}$*-algebra; and the Steenrod squares* (5.12) *satisfy*

$$(Sq^k\alpha)\overline{\lambda} = Sq^k(\alpha\cdot V(\overline{\lambda}))$$

for all $\overline{\lambda}\in\overline{\Lambda}$, *all* $\alpha\in\mathrm{Ext}_\Gamma^{*,*}(Q,N)$ *and all* $k\geq 0$.

As in Proposition 4.33 we note that the action of $\overline{\Lambda}$ passes to a quotient action of $\overline{\Omega}$. So Propostion 5.3 immediately implies:

PROPOSITION 5.4. *With the product* (5.11), $\mathrm{Ext}_\Gamma^{*,*}(Q,N)$ *becomes a commutative right* $\overline{\Omega}$*-algebra; and the Steenrod squares* (5.12) *satisfy*

$$(Sq^k\alpha)\overline{\omega} = Sq^k(\alpha\cdot V(\overline{\omega})) \tag{5.13}$$

for all $\overline{\omega}\in\overline{\Omega}$, *all* $\alpha\in\mathrm{Ext}_\Gamma^{*,*}(Q,N)$ *and all* $k\geq 0$.

Passing from vector spaces to their negatives we obtain a product

$$\mu : \overline{\mathrm{Ext}}^{\Gamma}_{*,*}(Q,N)\otimes\overline{\mathrm{Ext}}^{\Gamma}_{*,*}(Q,N)\to\overline{\mathrm{Ext}}^{\Gamma}_{*,*}(Q,N) \tag{5.14}$$

and squaring operations

$$\overline{Sq^k} : \overline{\mathrm{Ext}}^{\Gamma}_{-q,s}(Q,N)\to\overline{\mathrm{Ext}}^{\Gamma}_{-q-k,2s}(Q,N) \tag{5.15}$$

for all $k,q\geq 0$, $s\in\mathbb{S}$.

Recalling from Definition 1.70 the notion of the negative $\overline{\boldsymbol{\mathcal{H}}}$ of the algebra with coproducts $\boldsymbol{\mathcal{H}}$, we have from Remark 1.71:

PROPOSITION 5.5. *The operations* (5.14) *and* (5.15) *give* $\overline{\mathrm{Ext}}^{\Gamma}_{*,*}(Q,N)$ *the structure of a left* $\overline{\boldsymbol{\mathcal{H}}}$*-algebra.*

We can also take the negative of the right action of $\overline{\Lambda}$ on $\mathrm{Ext}_\Gamma^{q,*}(Q,N)$, obtaining a right action of Λ on $\overline{\mathrm{Ext}}^{\Gamma}_{-q,*}(Q,N)$ for each $q\geq 0$. Simultaneously we pass from the action of Λ to the quotient action of Ω, as in (4.38). Proposition 5.3 becomes:

PROPOSITION 5.6. *With the product* (5.14), $\overline{\mathrm{Ext}}^{\Gamma}_{*,*}(Q,N)$ *becomes a commutative right* Ω*-algebra, and the Steenrod squares* (5.15) *satisfy*

$$(\overline{Sq^k}\alpha)\omega = \overline{Sq^k}(\alpha\cdot V(\omega)) \tag{5.16}$$

for all $\omega\in\Omega$, *all* $\alpha\in\overline{\mathrm{Ext}}^{\Gamma}_{*,*}(Q,N)$ *and all* $k\geq 0$.

We immediately apply our result that $\overline{\mathrm{Ext}}^{\Gamma}_{*,*}(Q,N)$ is a commutative Ω-algebra, by giving the promised definitions of the operations (5.9) and (5.10b) upon the E_2-term of the change-of-rings spectral sequence.

DEFINITION 5.7. By the product (5.9) and diagonal Steenrod squares (5.10b) we mean the product and Steenrod squares that are ordinarily defined on the cohomology of the Hopf algebra Ω with coefficients in the Ω-coalgebra P and commutative Ω-algebra $\overline{\mathrm{Ext}}^{\Gamma}_{*,*}(Q,N)$, as in (1.125) and (1.126).

REMARK 5.8. We will explain how the gradings work in the above Definition. $\overline{\mathrm{Ext}}^{\Gamma}_{*,*}(Q,N)$ is graded on $\mathbb{Z}\times\mathbb{S}$; as are Ω and P: the latter two are concentrated in degrees $0\times\mathbb{S}$. Thus the role of "internal degree", the degree called $\mathbb{S}$ in (1.126), is here played by $\mathbb{Z}\times\mathbb{S}$, and the diagonal squares we have just defined treat the gradings as advertised in (5.10b).

As an immediate consequence of Proposition 1.120 we have:

PROPOSITION 5.9. *The diagonal Steenrod squares* (5.10b) *satisfy the Adem relations, and satisfy the Cartan formula with respect to the product* (5.9).

To define the vertical squares we are going to use the theory developed in Chapter 3, Section 2. The algebra with coproducts Π of that section we take here to be $\overline{\mathcal{H}}$. We begin by showing how to make any $\mathbb{S}$-graded coalgebra M over $\mathbb{Z}/2$ into a right module over $\overline{\mathcal{H}}$ in the sense of Definition 1.49. That definition requires that M be graded on an abelian group of the form $\mathbb{Z} \times \mathbb{S}$, where the first factor gives the "additive" degree, and the second the "multiplicative". We do this by concentrating M in additive degree zero:

$$M_{k,s} = \begin{cases} M_s & \text{if } k = 0 \\ 0 & \text{if } k > 0 \end{cases} \tag{5.17}$$

for all $s \in \mathbb{S}$. We define right actions of the Steenrod squares on M by setting

$$m\overline{Sq^k} = \begin{cases} V(m) & \text{if } k = 0 \\ 0 & \text{if } k > 0 \end{cases} \tag{5.18}$$

for all $m \in M$, where V is the Verschiebung. (5.18) is clearly compatible with the Adem relations (1.116), and so extends to a right action of $\overline{\mathcal{H}}$ on M. This action satisfies: $M_{0,s} \otimes \overline{\mathcal{H}}_{0,v} \to M_{0,s/2^v}$, as required by Definition 1.49.

DEFINITION 5.10. A right $\overline{\mathcal{H}}$-module M is called special if it is constructed from an $\mathbb{S}$-graded coalgebra M according to equations (5.17) and (5.18).

From (1.19) and the fact that $\psi(\overline{Sq^0}) = \overline{Sq^0} \otimes \overline{Sq^0}$ in $\overline{\mathcal{H}}$ we have:

PROPOSITION 5.11. *Let M and M' be coalgebras, both regarded as special right $\overline{\mathcal{H}}$-modules. If the tensor product $M \otimes M'$ of coalgebras is also regarded as a special right $\overline{\mathcal{H}}$-module, then it is isomorphic to the tensor product of the $\overline{\mathcal{H}}$-modules M and M'.*

Right modules over the algebra with coproducts $\overline{\mathcal{H}}$ can have extra structure, as described in Remark 1.54. In the case of special right $\overline{\mathcal{H}}$-modules we have from (1.18), (1.27) and (1.28):

PROPOSITION 5.12. *Let M be an $\mathbb{S}$-graded coalgebra, regarded as a special right $\overline{\mathcal{H}}$-module. Then:*
1. *M becomes a right $\overline{\mathcal{H}}$-coalgebra;*
2. *if M is a bialgebra, M becomes a right $\overline{\mathcal{H}}$-bialgebra;*
3. *if M is a Hopf algebra, M becomes a right $\overline{\mathcal{H}}$-Hopf algebra.*

In the same spirit we have:

PROPOSITION 5.13. *Let Ω be a bialgebra, and M a right Ω-coalgebra. If we regard both Ω and M as special right $\overline{\mathcal{H}}$-modules, then Ω becomes a right $\overline{\mathcal{H}}$-algebra, and M becomes a right $\overline{\mathcal{H}} - \Omega$ module.*

PROOF. It remains only to check that the action $\sigma_M : M \otimes \Omega \to M$ is $\overline{\mathcal{H}}$-linear. But this follows immediately from Proposition 1.30. □

Recall the notion of a tensor product of Ω-coalgebras from Proposition 1.29.

PROPOSITION 5.14. *Let Ω be a Hopf algebra, regarded as a right $\overline{\mathcal{H}}$-Hopf algebra under the special action of $\overline{\mathcal{H}}$. Let M be a right Ω-coalgebra, and let the tensor product $M \otimes \Omega$ of the Ω-coalgebras M and Ω be regarded as a special right $\overline{\mathcal{H}}$-module. Then $M \otimes \Omega$ is a relative projective in the category of $\overline{\mathcal{H}} - \Omega$ modules.*

PROOF. This follows immediately from Propositions 5.11 and 1.66. □

REMARK 5.15. If Ω is a Hopf algebra and P a right Ω-coalgebra, we can, as in Example 1.109, form the simplicial object $W^{\Omega}(P)$ over the category of Ω-coalgebras, augmenting to P. We then regard $W^{\Omega}_p(P)$ as an $\overline{\mathcal{H}} - \Omega$ module under the special $\overline{\mathcal{H}}$-action, for each $p \geq -1$. We claim that the associated chain complex $CW^{\Omega}(P)$ is a resolution of P by relative projectives in the category of $\overline{\mathcal{H}} - \Omega$ modules, in the sense of Definition 1.68. In fact this follows from Proposition 5.14, together with the fact that the contracting homotopies $h : W^{\Omega}_{p-1}(P) \to W^{\Omega}_p(P)$ described in (1.98) are in the present context homomorphisms of coalgebras, and consequently homomorphisms of right $\overline{\mathcal{H}}$-modules.

Now we return to the situation described at the beginning of this chapter: we have an extension of Hopf algebras (5.2), a right Ω-coalgebra P, a right Λ-coalgebra Q, and a commutative right Λ-algebra N. If we regard Ω as a special right $\overline{\mathcal{H}}$-module then, according to Proposition 5.12, Ω becomes a right $\overline{\mathcal{H}}$-Hopf algebra. So we can make a nice interpretation of the second part of Proposition 5.6.

PROPOSITION 5.16. *The left action of $\mathcal{H}$ upon $\overline{\mathrm{Ext}}^{\Gamma}_{*,*}(Q, N)$, as defined by (5.15), is compatible with the right action of Ω, in the sense of Definition 3.25.*

PROOF. Equation (3.4) reduces to (5.16) in the present context. □

On the other hand, if we regard both Ω and P as special right $\overline{\mathcal{H}}$-modules then by Proposition 5.13, P becomes a right $\overline{\mathcal{H}} - \Omega$ module. So we can apply the theory of Chapter 3, Section 2, particularly Definition 3.31, with the substitutions:

$$\mathbb{T} \to \mathbb{Z}, \quad \Pi \to \overline{\mathcal{H}}, \quad \Xi \to \Omega, \quad M \to P, \quad N \to \overline{\mathrm{Ext}}^{\Gamma}_{*,*}(Q, N). \tag{5.19}$$

As in Remark 5.15, we can choose a resolution $\mathcal{P}(P)$ of P by relative projectives in the category of $\overline{\mathcal{H}} - \Omega$ modules, having the property that each $\mathcal{P}_p(P)$ is special as a module over $\overline{\mathcal{H}}$. Then from Proposition 3.28 we find that for each $p \geq 0$, $\mathrm{Hom}_{\Omega}(\mathcal{P}_p(P), \overline{\mathrm{Ext}}^{\Gamma}_{*,*}(Q, N))$ becomes a left $\mathcal{H}$-module, with action defined by (3.1). But if we take $\pi = Sq^k$ on the left of (3.1), only one term survives on the right, since $\overline{Sq^i}$ acts non-trivially on a special $\overline{\mathcal{H}}$-module only if $i = 0$. Then we have as in Definition 3.31:

PROPOSITION 5.17. *Let $\mathcal{P}(P)$ be a resolution of P by relative projectives in the category of $\overline{\mathcal{H}} - \Omega$ modules, for which each $\mathcal{P}_p(P)$ is special as a $\overline{\mathcal{H}}$-module. Then for each $p \geq 0$, the $\mathbb{Z} \times \mathbb{S}$-graded vector space $\mathrm{Hom}_{\Omega}(\mathcal{P}_p(P), \overline{\mathrm{Ext}}^{\Gamma}_{*,*}(Q, N))$ becomes a left $\mathcal{H}$-module, with action defined by*

$$(Sq^k \cdot f)(m) = \overline{Sq^k} f(m\overline{Sq^0}) \tag{5.20}$$

for all $f \in \mathrm{Hom}_{\Omega}(\mathcal{P}_p(P), \overline{\mathrm{Ext}}^{\Gamma}_{*,*}(Q, N))$, $m \in \mathcal{P}_p(P)$, and all $k \geq 0$. We have further that $\mathrm{Hom}_{\Omega}(\mathcal{P}(P), \overline{\mathrm{Ext}}^{\Gamma}_{*,*}(Q, N))$ is a cochain complex over the category of *left $\mathcal{H}$-modules, so the equation*

$$\mathrm{Ext}_\Omega^p(P, \overline{\mathrm{Ext}}_{*,*}^\Gamma(Q,N)) = H^p \operatorname{Hom}_\Omega(\mathcal{P}(P), \overline{\mathrm{Ext}}_{*,*}^\Gamma(Q,N)) \tag{5.21}$$

defines $\mathrm{Ext}_\Omega^p(P, \overline{\mathrm{Ext}}_{*,*}^\Gamma(Q,N))$ *as a left $\mathcal{H}$-module, for each $p \geq 0$.*

REMARK 5.18. Since each $\mathcal{P}_p(P)$ is "special" as a $\overline{\mathcal{H}}$-module it is concentrated in additive degree $q = 0$, so the convention (1.3a) allows us to write for all $p, q \geq 0$, $s \in \mathbb{S}$:

$$\operatorname{Hom}_\Omega^{q,s}(\mathcal{P}_p(P), \overline{\mathrm{Ext}}_{*,*}^\Gamma(Q,N)) = \operatorname{Hom}_\Omega^{s}(\mathcal{P}_p(P), \overline{\mathrm{Ext}}_{-q,*}^\Gamma(Q,N)).$$

Therefore one can write

$$\mathrm{Ext}_\Omega^{p,q,s}(P, \overline{\mathrm{Ext}}_{*,*}^\Gamma(Q,N)) = \mathrm{Ext}_\Omega^{p,s}(P, \overline{\mathrm{Ext}}_{-q}^\Gamma(Q,N)) \tag{5.22}$$

where on the right-hand side we have reverted to our old practice of suppressing internal degree from the notation for $\overline{\mathrm{Ext}}^\Gamma$. The notation on the right-hand side is the one we have used in setting up the change-of-rings spectral sequence. It emphasizes the fact that $\overline{\mathrm{Ext}}_{-q}^\Gamma(Q,N)$ is an Ω-module for each $q \geq 0$. The notation on the left-hand side is consistent with that of Chapter 3, Section 2. In either case an element in the vector space (5.22) has additive degree q and multiplicative degree s with respect to the action of the algebra with coproducts $\mathcal{H}$.

DEFINITION 5.19. By the vertical Steenrod operations (5.10a) we mean those defined by the left $\mathcal{H}$-action on $\mathrm{Ext}_\Omega^{p,*}(P, \overline{\mathrm{Ext}}_*^\Gamma(Q,N))$ as given by (5.20), (5.21), and (5.22).

REMARK 5.20. Since Sq^k has additive degree k and multiplicative degree 1 in the algebra with coproducts $\mathcal{H}$, it is clear that our vertical squares treat the gradings as advertised in (5.10a).

REMARK 5.21. We have seen in Remark 5.15 that one can choose the resolution $\mathcal{P}(P)$ in such a way that $\mathcal{P}_p(P)$ is special as a $\overline{\mathcal{H}}$-module for each $p \geq 0$. In this case each $\mathcal{P}_p(P)$ will be a coalgebra, and the formula (5.20) reduces to

$$(Sq^k \cdot f)(m) = \overline{Sq^k} f(Vm) \tag{5.23}$$

for all $f \in \operatorname{Hom}_\Omega(\mathcal{P}_p(P), \overline{\mathrm{Ext}}_{*,*}^\Gamma(Q,N))$, $m \in \mathcal{P}_p(P)$, and all $k \geq 0$.

We record the fact that the vertical squares come from an action of $\mathcal{H}$:

PROPOSITION 5.22. *The vertical squaring operations satisfy the Adem relations.*

We make further use of the theory in Chapter 3 to deduce some properties of these vertical squares. We apply Proposition 3.32 with the replacements (5.19). That the hypotheses of that proposition are satisfied is a consequence of Propositions 5.5, 5.6, 5.11, 5.12, 5.13. We conclude:

PROPOSITION 5.23. *The vertical Steenrod squares on $\mathrm{Ext}_\Omega^{*,*}(P, \overline{\mathrm{Ext}}_*^\Gamma(Q,N))$ satisfy the Cartan formula. They are related to the diagonal squaring operations (5.10b), as given by Definition 5.7, by the equation*

$$Sq_V^l Sq_D^k = Sq_D^k Sq_V^{l/2}$$

for all $k, l \geq 0$.

3. Central Extensions

We consider the case in which Γ is central as a sub-algebra of Λ, and both Q and N are trivial as right Λ-modules. This is the case considered by Uehara, [**86**]. From Proposition 4.34 we have that Ω acts trivially on $\overline{\mathrm{Ext}}^{\Gamma}_{-q}(Q,N)$ for each $q \geq 0$. We assume in addition:

1. both Ω and P are locally finite and bounded below;
2. $\mathrm{Ext}^q_\Gamma(Q,N)$ is bounded below for each $q \geq 0$.

Then $\overline{\mathrm{Ext}}^{\Gamma}_{-q}(Q,N)$ is bounded above for each $q \geq 0$ and we have, as in (1.133), an isomorphism of $\mathbb{S}$-graded vector spaces:

$$\mathrm{Ext}^{p,*}_\Omega(P, \overline{\mathrm{Ext}}^{\Gamma}_{-q}(Q,N)) = \mathrm{Ext}^{p,*}_\Omega(P,\mathbb{Z}/2) \otimes \mathrm{Ext}^{q,*}_\Gamma(Q,N). \tag{5.24}$$

PROPOSITION 5.24. *Suppose* (5.2) *is an extension of Hopf algebras for which* Γ *is a central sub-algebra of* Λ. *Suppose* P *is a right* Ω*-coalgebra. Suppose that* Q *is a coalgebra and that* N *is a commutative algebra; both regarded as trivial right* Λ*-modules. Suppose finally that conditions 1 and 2 listed above are satisfied. Then* (5.24) *is an isomorphism of algebras, and the operations* (5.10) *are given by*

$$Sq^k_V(u \otimes v) = Sq^0 u \otimes Sq^k v \tag{5.25a}$$

$$Sq^k_D(u \otimes v) = Sq^k u \otimes v^2 \tag{5.25b}$$

for all $u \in \mathrm{Ext}^{p,*}_\Omega(P,F_2)$ *and all* $v \in \mathrm{Ext}^{q,*}_\Gamma(Q,N)$.

PROOF. The statements that (5.24) is an isomorphism of algebras, and that the diagonal squares are given by (5.25b), are implied by Remark 1.122. We will show that the vertical squares are given by (5.25a). We begin by observing that for each fixed $q \geq 0$ the vertical square (5.10a) is operating on the homology groups of the complex $\mathrm{Hom}_\Omega(CW^\Omega(P), \overline{\mathrm{Ext}}^{\Gamma}_{-q}(Q,N))$, which under the current hypotheses can be identified with the complex $\mathrm{Hom}_\Omega(CW^\Omega(P),\mathbb{Z}/2) \otimes \mathrm{Ext}^q_\Gamma(Q,N)$. Then the cochain operation (5.23) that induces Sq^k_V is identified with an operation:

$$Sq^k : \mathrm{Hom}_\Omega(CW^\Omega(P),\mathbb{Z}/2) \otimes \mathrm{Ext}^q_\Gamma(Q,N) \to \mathrm{Hom}_\Omega(CW^\Omega(P),\mathbb{Z}/2) \otimes \mathrm{Ext}^{q+k}_\Gamma(Q,N)$$

that can be described as follows. Given $g \in \mathrm{Hom}_\Omega(CW^\Omega(P),\mathbb{Z}/2)$ and given an element $v \in \mathrm{Ext}^q_\Gamma(Q,N)$, then

$$Sq^k(g \otimes v) = \overline{g} \otimes Sq^k v$$

where $\overline{g} \in \mathrm{Hom}_\Omega(CW^\Omega(P),\mathbb{Z}/2)$ is defined by

$$\overline{g}(m) = g(Vm)$$

for all $m \in CW^\Omega(P)$. So if g is a cocycle representing a class $u \in \mathrm{Ext}^{p,*}_\Omega(P,\mathbb{Z}/2)$, Proposition 1.114 shows that $\overline{g}$ represents $Sq^0 u$. □

4. The Operations at the E_2-level

Throughout this section we fix an extension of Hopf algebras (5.2) over $\mathbb{Z}/2$. We fix a right Ω-coalgebra P, a right Λ-coalgebra Q, and a commutative right Λ-algebra N. We study the change-of-rings spectral sequence (5.1). Our aim is to show that if the E_2-term is identified as in (5.1), then the product and squaring operations at the E_2-level are those that were described in purely Hopf-theoretic terms in Section 2. In case the extension (5.2) is central and $P = Q = N = \mathbb{Z}/2$ the descriptions of Section 3 apply, and our results reduce to those of Uehara [**86**].

Internal grading plays no essential role in this section; we have suppressed it.

We begin with some conventions. We write X for the bisimplicial Λ-coalgebra given by (5.3), and $\{E_r, d_r\}$ for the spectral sequence of this bisimplicial Λ-coalgebra with coefficients in N. Suppose an element $\gamma \in E_2^{m,n}$ is represented in the sense of (1.149) by a Λ-linear cochain $z : (X_{m,n} = W_m^{\Omega}(P) \otimes W_n^{\Lambda}(Q)) \to N$. Then we will write $\tilde{z} : W_m^{\Omega}(P) \to \overline{\mathrm{Hom}}^{\Gamma}(W_n^{\Lambda}(Q), N)$ for the Ω-linear mapping adjoint to z under (4.46). Since in the representation (1.149) one has always $\delta^v z = 0$, one knows that $\tilde{z}$ maps into the cocycles of the cochain complex $\overline{\mathrm{Hom}}^{\Gamma}(CW^{\Lambda}(Q), N)$, and so defines an Ω-linear mapping $\tilde{\tilde{z}} : W_m^{\Omega}(P) \to \overline{\mathrm{Ext}}_n^{\Gamma}(Q, N)$. Then γ, viewed as an element of $\mathrm{Ext}_{\Omega}^m(P, \overline{\mathrm{Ext}}_n^{\Gamma}(Q, N))$, is characterized by the corresponding $\tilde{\tilde{z}}$. We will use these facts and notations throughout the proofs of the next three propositions.

PROPOSITION 5.25. *The product at the E_2-level of the change-of-rings spectral sequence is the product in Hopf algebra cohomology described by equation* (5.9) *and Definition 5.7.*

PROOF. Suppose given $\alpha \in E_2^{p,q}$ and $\beta \in E_2^{r,s}$; and suppose these are represented in the sense of (1.149) by Λ-linear cochains $x : W_p^{\Omega}(P) \otimes W_q^{\Lambda}(Q) \to N$, $y : W_r^{\Omega}(P) \otimes W_s^{\Lambda}(Q) \to N$, respectively. Using (2.59) and Proposition 1.92 we get a representation of $\alpha \cdot \beta$ in $E_2^{p+r,q+s}$ by a cochain $z : W_{p+r}^{\Omega}(P) \otimes W_{q+s}^{\Lambda}(Q) \to N$. This cochain is the following composition:

(5.26)

$$W_{p+r}^{\Omega}(P) \otimes W_{q+s}^{\Lambda}(Q) \xrightarrow{\psi\otimes\psi} W_{p+r}^{\Omega}(P) \otimes W_{p+r}^{\Omega}(P) \otimes W_{q+s}^{\Lambda}(Q) \otimes W_{q+s}^{\Lambda}(Q) -$$
$$\xrightarrow{D_0^{p,r}(W^{\Omega}(P)) \otimes D_0^{q,s}(W^{\Lambda}(Q))} W_p^{\Omega}(P) \otimes W_r^{\Omega}(P) \otimes W_q^{\Lambda}(Q) \otimes W_s^{\Lambda}(Q) -$$
$$\xrightarrow{(1,3,2,4)} W_p^{\Omega}(P) \otimes W_q^{\Lambda}(Q) \otimes W_r^{\Omega}(P) \otimes W_s^{\Lambda}(Q) \xrightarrow{x\otimes y} N \otimes N \xrightarrow{\mu} N.$$

We must show that the associated Ω-linear mapping $\tilde{\tilde{z}} : W_{p+r}^{\Omega}(P) \to \overline{\mathrm{Ext}}_{q+s}^{\Gamma}(Q, N)$ represents the same element in $\mathrm{Ext}_{\Omega}^{p+r}(P, \overline{\mathrm{Ext}}_{q+s}^{\Gamma}(Q, N))$ as the product of the two classes $\alpha \in \mathrm{Ext}_{\Omega}^p(P, \overline{\mathrm{Ext}}_q^{\Gamma}(Q, N))$, $\beta \in \mathrm{Ext}_{\Omega}^r(P, \overline{\mathrm{Ext}}_s^{\Gamma}(Q, N))$ as calculated in $\mathrm{Ext}_{\Omega}^*(P, \overline{\mathrm{Ext}}_*^{\Gamma}(Q, N))$. But $\tilde{\tilde{z}}$ is represented by $\tilde{z} : W_{p+r}^{\Omega}(P) \to \overline{\mathrm{Hom}}^{\Gamma}(W_{q+s}^{\Lambda}(Q), N)$, and one reads off from (5.26) that for each $w \in W_{p+r}^{\Omega}(P)$, $\tilde{z}(w) : W_{q+s}^{\Lambda}(Q) \to N$ is the following Γ-linear composition:

$$(5.27) \qquad W_{q+s}^{\Lambda}(Q) \xrightarrow{\psi} W_{q+s}^{\Lambda}(Q) \otimes W_{q+s}^{\Lambda}(Q) \xrightarrow{D_0^{q,s}(W^{\Lambda}(Q))} W_q^{\Lambda}(Q) \otimes W_s^{\Lambda}(Q) -$$
$$\xrightarrow{\sum_i \tilde{x}(w_i)\otimes\tilde{y}(w_i')} N \otimes N \xrightarrow{\mu} N.$$

Here we have written $\sum_i w_i \otimes w_i'$ for the image of w under the composition

$$(5.28) \qquad W_{p+r}^{\Omega}(P) \xrightarrow{\psi} W_{p+r}^{\Omega}(P) \otimes W_{p+r}^{\Omega}(P) \xrightarrow{D_0^{p,r}(W^{\Omega}(P))} W_p^{\Omega}(P) \otimes W_r^{\Omega}(P).$$

One reads equation (5.27) as saying that for each $w \in W_{p+r}^{\Omega}(P)$,

$$\tilde{\tilde{z}}(w) = \sum_i \tilde{\tilde{x}}(w_i) \cdot \tilde{\tilde{y}}(w_i')$$

where the product on the right-hand side is computed in $\overline{\mathrm{Ext}}^{\Gamma}_{*}(Q,N)$. It follows that the mapping $\tilde{\tilde{z}}$ is just the following composition:

$$W^{\Omega}_{p+r}(P) \xrightarrow{\psi} W^{\Omega}_{p+r}(P) \otimes W^{\Omega}_{p+r}(P) \xrightarrow{D_0^{p,r}(W^{\Omega}(P))} W^{\Omega}_{p}(P) \otimes W^{\Omega}_{r}(P) \text{—}$$
$$\xrightarrow{\tilde{\tilde{x}}\otimes\tilde{\tilde{y}}} \overline{\mathrm{Ext}}^{\Gamma}_{q}(Q,N) \otimes \overline{\mathrm{Ext}}^{\Gamma}_{s}(Q,N) \xrightarrow{\mu} \overline{\mathrm{Ext}}^{\Gamma}_{q+s}(Q,N).$$

But this Ω-linear cochain indeed represents the product $\alpha{\cdot}\beta$ in $\mathrm{Ext}^{*}_{\Omega}(P, \overline{\mathrm{Ext}}^{\Gamma}_{*}(Q,N))$, as required. □

PROPOSITION 5.26. *The diagonal Steenrod operations* $Sq^k_D : E_2^{p,q} \to E_2^{p+k,2q}$ *in the change-of-rings spectral sequence agree with the diagonal operations on Hopf algebra cohomology described in equation* (5.10b) *and Definition 5.7.*

PROOF. Let $\alpha \in E_2^{p,q} = \mathrm{Ext}^{p}_{\Omega}(P, \overline{\mathrm{Ext}}^{\Gamma}_{-q}(Q,N))$ be given. Then we have two versions of $Sq^k_D\alpha$ in $E_2^{p+k,2q} = \mathrm{Ext}^{p+k}_{\Omega}(P, \overline{\mathrm{Ext}}^{\Gamma}_{-2q}(Q,N))$ ... one described by (2.61) and the other by Definition 5.7. We will represent both as cocycles in the cochain complex $\mathrm{Hom}_{\Omega}(CW^{\Omega}(P), \overline{\mathrm{Ext}}^{\Gamma}_{-2q}(Q,N))$, and we will show these cocyles equal. Choose a Λ-linear cochain $x : X_{p,q} \to N$ that represents α in the sense of (1.149). Then by (2.61), $Sq^k_D\alpha$ is represented by a Λ-linear cochain $y : X_{p+k,2q} \to N$ that we can write, using Proposition 1.92, as the following composition:

$$W^{\Omega}_{p+k}(P) \otimes W^{\Lambda}_{2q}(Q) \xrightarrow{\psi\otimes\psi} W^{\Omega}_{p+k}(P) \otimes W^{\Omega}_{p+k}(P) \otimes W^{\Lambda}_{2q}(Q) \otimes W^{\Lambda}_{2q}(Q) \text{——}$$
$$\xrightarrow{D^{p,p}_{p-k}\otimes D^{q,q}_{0}} W^{\Omega}_{p}(P) \otimes W^{\Omega}_{p}(P) \otimes W^{\Lambda}_{q}(Q) \otimes W^{\Lambda}_{q}(Q) \xrightarrow{(1,3,2,4)}$$
$$W^{\Omega}_{p}(P) \otimes W^{\Lambda}_{q}(Q) \otimes W^{\Omega}_{p}(P) \otimes W^{\Lambda}_{q}(Q) \xrightarrow{x\otimes x} N \otimes N \xrightarrow{\mu} N.$$

An Ω-linear mapping $\tilde{y} : W^{\Omega}_{p+k}(P) \to \overline{\mathrm{Hom}}^{\Gamma}(W^{\Lambda}_{2q}(Q), N)$ is adjoint to y under the isomorphism (4.46). It is given by the following composition:

(5.29)

$$W^{\Omega}_{p+k}(P) \xrightarrow{\psi} W^{\Omega}_{p+k}(P) \otimes W^{\Omega}_{p+k}(P) \xrightarrow{D^{p,p}_{p-k}} W^{\Omega}_{p}(P) \otimes W^{\Omega}_{p}(P) \xrightarrow{\tilde{x}\otimes\tilde{x}}$$
$$\overline{\mathrm{Hom}}^{\Gamma}(C_qW^{\Lambda}(Q), N) \otimes \overline{\mathrm{Hom}}^{\Gamma}(C_qW^{\Lambda}(Q), N) \xrightarrow{\phi} \overline{\mathrm{Hom}}^{\Gamma}((CW^{\Lambda}(Q) \otimes CW^{\Lambda}(Q))_{2q}, N)$$
$$\xrightarrow{D_0^*} \overline{\mathrm{Hom}}^{\Gamma}(C_{2q}(W^{\Lambda}(Q) \times W^{\Lambda}(Q)), N) \xrightarrow{\psi^*} \overline{\mathrm{Hom}}^{\Gamma}(C_{2q}W^{\Lambda}(Q), N),$$

where ϕ is as in (1.129) and $\tilde{x}$ is adjoint to x. We note that the cochain $\tilde{x}$ represents an Ω-linear mapping $\tilde{\tilde{x}} : W^{\Omega}_{p}(P) \to \overline{\mathrm{Ext}}^{\Gamma}_{-q}(Q,N)$. Hence (5.29) represents an Ω-linear homomorphism $\tilde{\tilde{y}} : W^{\Omega}_{p+k}(P) \to \overline{\mathrm{Ext}}^{\Gamma}_{-2q}(Q,N)$. This is the following composition:

$$(5.30) \qquad W^{\Omega}_{p+k}(P) \xrightarrow{\psi} W^{\Omega}_{p+k}(P) \otimes W^{\Omega}_{p+k}(P) \xrightarrow{D^{p,p}_{p-k}} W^{\Omega}_{p}(P) \otimes W^{\Omega}_{p}(P) \text{—}$$
$$\xrightarrow{\tilde{\tilde{x}}\otimes\tilde{\tilde{x}}} \overline{\mathrm{Ext}}^{\Gamma}_{-q}(Q,N) \otimes \overline{\mathrm{Ext}}^{\Gamma}_{-q}(Q,N) \xrightarrow{\mu} \overline{\mathrm{Ext}}^{\Gamma}_{-2q}(Q,N).$$

Here μ is the (negative of the) product in Hopf algebra cohomology as defined by (1.125). But $\tilde{\tilde{x}}$ represents the class $\alpha \in \operatorname{Ext}^p_\Omega(P, \overline{\operatorname{Ext}}^\Gamma_{-q}(Q,N))$, so it is clear that the composition (5.30) represents $Sq^k_D\alpha \in \operatorname{Ext}^{p+k}_\Omega(P, \overline{\operatorname{Ext}}^\Gamma_{-2q}(Q,N))$ as given by Definition 5.7. This observation completes our proof. □

PROPOSITION 5.27. *The vertical Steenrod operations $Sq^k_V : E_2^{p,q} \to E_2^{p,q+k}$ in the change-of-rings spectral sequence agree with the vertical operations described by equation* (5.10a) *and Definition 5.19.*

PROOF. Let $\alpha \in E_2^{p,q} = \operatorname{Ext}^p_\Omega(P, \overline{\operatorname{Ext}}^\Gamma_{-q}(Q,N))$ be given. Then we have two versions of $Sq^k_V\alpha$ in $E_2^{p,q+k} = \operatorname{Ext}^p_\Omega(P, \overline{\operatorname{Ext}}^\Gamma_{-q-k}(Q,N))$: one described by (2.60) and the other by Definition 5.19. We will represent both as cocycles in the cochain complex $\operatorname{Hom}_\Omega(CW^\Omega(P), \overline{\operatorname{Ext}}^\Gamma_{-q-k}(Q,N))$, and we will show these cocyles equal. With the bisimplicial Λ-coalgebra X as given in (5.3), we choose a Λ-linear cochain $x : X_{p,q} \to N$ that represents α in the sense of (1.149). Then $Sq^k_V\alpha$ is represented by the Λ-linear cochain (2.60), which we call $y : X_{p,q+k} \to N$. Using Proposition 1.92 we can write y as the following composition:

$$\begin{aligned}(5.31)\quad & W^\Omega_p(P)\otimes W^\Lambda_{q+k}(Q) \xrightarrow{\psi\otimes\psi} W^\Omega_p(P)\otimes W^\Omega_p(P)\otimes W^\Lambda_{q+k}(Q)\otimes W^\Lambda_{q+k}(Q) \text{——} \\ & \xrightarrow{W^\Omega_p(P)\otimes W^\Omega_p(P)\otimes D^{q,q}_{q-k}} W^\Omega_p(P)\otimes W^\Omega_p(P)\otimes W^\Lambda_q(Q)\otimes W^\Lambda_q(Q) \text{——} \\ & \xrightarrow{(1,3,2,4)} W^\Omega_p(P)\otimes W^\Lambda_q(Q)\otimes W^\Omega_p(P)\otimes W^\Lambda_q(Q) \xrightarrow{x\otimes x} N\otimes N \xrightarrow{\mu} N.\end{aligned}$$

Let $\tilde{y} : W^\Omega_p(P) \to \overline{\operatorname{Hom}}^\Gamma(W^\Lambda_{q+k}(Q), N)$ be the Ω-linear homomorphism adjoint to y under (4.46). We wish to describe for each $w \in W^\Omega_p(P)$ the Γ-linear homomorphism: $\tilde{y}(w) : W^\Lambda_{q+k}(Q) \to N$. As in (1.17) we expand

$$\psi(w) = \sum_i (w_i \otimes w_i' + w_i' \otimes w_i) + V(w)\otimes V(w),$$

where V is the Verschiebung. Then from (5.31) we have that $\tilde{y}(w)$ is the composition

$$\begin{aligned}(5.32)\quad & W^\Lambda_{q+k}(Q) \xrightarrow{\psi} W^\Lambda_{q+k}(Q)\otimes W^\Lambda_{q+k}(Q) \xrightarrow{D^{q,q}_{q-k}} W^\Lambda_q(Q)\otimes W^\Lambda_q(Q) \text{——} \\ & \xrightarrow{\tilde{x}V(w)\otimes\tilde{x}V(w)+\sum_i[\tilde{x}(w_i)\otimes\tilde{x}(w_i')+\tilde{x}(w_i')\otimes\tilde{x}(w_i)]} N\otimes N \xrightarrow{\mu} N.\end{aligned}$$

Here we have written $\tilde{x} : W^\Omega_p(P) \to \overline{\operatorname{Hom}}^\Gamma(W^\Lambda_q(Q), N)$ for the Ω-linear mapping adjoint to x. (5.32) defines in turn an element $\tilde{\tilde{y}}(w) \in \overline{\operatorname{Ext}}^\Gamma_{-q-k}(Q,N)$, and the function $\tilde{\tilde{y}} : W^\Omega_p(P) \to \overline{\operatorname{Ext}}^\Gamma_{-q-k}(Q,N)$ characterizes in $E_2^{p,q+k}$ the element $Sq^k_V\alpha$ in $\operatorname{Ext}^p_\Omega(P, \overline{\operatorname{Ext}}^\Gamma_{-q-k}(Q,N))$ as it has been defined by (2.60).

Now we consider $Sq^k_V\alpha$ as given by Defintion 5.19. We will represent this element also as an Ω-linear homomorphism from $W^\Omega_p(P)$ to $\overline{\operatorname{Ext}}^\Gamma_{-q-k}(Q,N)$. Considered as an element of $\operatorname{Ext}^p_\Omega(P, \overline{\operatorname{Ext}}^\Gamma_{-q}(Q,N))$, α is represented by the Ω-linear cocycle $\tilde{\tilde{x}} : C_pW^{\Omega}(P) \to \overline{\mathrm{Ext}}^{\Gamma}_{-q}(Q,N)$ that is defined by $\tilde{x}$ above. In Definition 5.19 we are regarding P as a special $\overline{\mathcal{H}}$-module, and as an $\overline{\mathcal{H}} - \Omega$ module. By Remark 5.15, $CW^{\Omega}(P)$ is a resolution of P by relative projectives in the category of $\overline{\mathcal{H}} - \Omega$ modules, and each projective is special as a module over $\overline{\mathcal{H}}$. It follows from Remark 5.21 that $Sq^k_V\alpha \in \mathrm{Ext}^p_{\Omega}(P, \overline{\mathrm{Ext}}^{\Gamma}_{-q-k}(Q,N))$ is represented by the following composition:

$$C_pW^{\Omega}(P) \xrightarrow{V} C_pW^{\Omega}(P) \xrightarrow{\tilde{\tilde{x}}} \overline{\mathrm{Ext}}^{\Gamma}_{-q}(Q,N) \xrightarrow{\overline{Sq}^k} \overline{\mathrm{Ext}}^{\Gamma}_{-q-k}(Q,N).$$

The $\overline{Sq^k}$ in this diagram is that defined by (5.15). We conclude that $Sq^k_V\alpha$ is represented by an Ω-linear homomorphism $W^{\Omega}_p(P) \to \overline{\mathrm{Hom}}^{\Gamma}(W^{\Lambda}_{q+k}(Q), N)$ whose value on an arbitrary $w \in W^{\Omega}_p(P)$ is the Γ-linear mapping

$$(5.33)\qquad W^{\Lambda}_{q+k}(Q) \xrightarrow{\psi} (W^{\Lambda}(Q) \times W^{\Lambda}(Q))_{q+k} \xrightarrow{D^{q,q}_{q-k}} W^{\Lambda}_q(Q) \otimes W^{\Lambda}_q(Q) \text{—}$$
$$\xrightarrow{\tilde{x}(V(w)) \otimes \tilde{x}(V(w))} N \otimes N \xrightarrow{\mu} N.$$

To complete the proof of our theorem, we need only show that the compositions (5.32) and (5.33) represent equal elements of $\overline{\mathrm{Ext}}^{\Gamma}_{-q-k}(Q,N)$. Thus, it will be enough to show that each term in (5.32) of the form

$$(5.34)\qquad \mu\left(\tilde{x}(w_i) \otimes \tilde{x}(w_i') + \tilde{x}(w_i') \otimes \tilde{x}(w_i)\right) D^{q,q}_{q-k}\psi$$

is a coboundary in the cochain complex $\overline{\mathrm{Hom}}^{\Gamma}(CW^{\Lambda}(Q), N)$. But by the commutativity of N and the cocommutativity of $W^{\Lambda}_{q+k}(Q)$ we have

$$\mu\left(\tilde{x}(w_i') \otimes \tilde{x}(w_i)\right)(D_{q-k})\psi = \mu\left(\tilde{x}(w_i) \otimes \tilde{x}(w_i')\right) TD_{q-k}T\psi.$$

Hence we can use (1.59) to write (5.34) as

(5.35)
$$\mu\left(\tilde{x}(w_i) \otimes \tilde{x}(w_i')\right)(D_{q-k} + TD_{q-k}T)\psi = \mu\left(\tilde{x}(w_i) \otimes \tilde{x}(w_i')\right)(\partial D_{q-k+1} + D_{q-k+1}\partial)\psi.$$

But $\delta^v x = 0$ so one knows (as in our comments at the beginning of this section), that $\tilde{x}(w_i)$ and $\tilde{x}(w_i')$ are cocycles in the cochain complex $\overline{\mathrm{Hom}}^{\Gamma}(CW^{\Lambda}(Q), N)$. Hence (5.35) reduces to the following composition:

$$\mu\left(\tilde{x}(w_i) \otimes \tilde{x}(w_i')\right) D_{q-k+1}\partial\psi = \mu\left(\tilde{x}(w_i) \otimes \tilde{x}(w_i')\right) D_{q-k+1}\psi\partial.$$

This is manifestly a coboundary in $\overline{\mathrm{Hom}}^{\Gamma}(CW^{\Lambda}(Q), N)$, so we are done. □

5. A Simple Example

We consider the spectral sequence for a particular extension of Hopf algebras. The extension is central, so the results of [**86**] are sufficient. We give some examples of actions of the Steenrod squares, but we do not work out the spectral sequence completely.

Consider the sequence of graded polynomial algebras

$$\mathbb{Z}/2[\xi_1] \to \mathbb{Z}/2[\xi_1, \xi_2] \to \mathbb{Z}/2[\xi_2]$$

with grading determined by $|\xi_1| = 1$, $|\xi_2| = 3$. We make this into a diagram of Hopf algebras by declaring the first and last to be primitively generated, while in the middle term setting $\psi(\xi_2) = \xi_2 \otimes 1 + \xi_1^2 \otimes \xi_1 + 1 \otimes \xi_2$. Taking duals we get an extension diagram of cocommuative Hopf algebras

$$\Gamma(Sq^{0,1}) \to \Lambda \to \Gamma(Sq^1) \tag{5.36}$$

where we are writing $\Gamma(\theta)$ for the algebra of divided powers on the element θ. Of course, both Λ and $\Gamma(Sq^1)$ are quotients of the Steenrod algebra $\boldsymbol{\mathcal{A}}$. One sees easily that the extension (5.36) is central. So by (4.49) and (5.24) the associated change-of-rings spectral sequence converging to $\mathrm{Ext}_\Lambda(\mathbb{Z}/2, \mathbb{Z}/2)$ has

$$\begin{aligned} E_2^{p,q} &= \mathrm{Ext}^p_{\Gamma(Sq^1)}(\mathbb{Z}/2, \mathbb{Z}/2) \otimes \mathrm{Ext}^q_{\Gamma(Sq^{0,1})}(\mathbb{Z}/2, \mathbb{Z}/2) \\ &= \mathbb{Z}/2[h_0, h_1, \dots] \otimes \mathbb{Z}/2[w_0, w_1, \dots]. \end{aligned}$$

Here for each $i \geq 0$, h_i is the class corresponding to $|\xi_1^{2^i}|$ in the cobar construction on $\mathbb{Z}/2[\xi_1]$, and w_i is the class corresponding to $|\xi_2^{2^i}|$ in the cobar construction on $\mathbb{Z}/2[\xi_2]$. The actions of $\boldsymbol{\mathcal{H}}$ on the cohomology rings $\mathrm{Ext}^{*,*}_{\Gamma(Sq^1)}(\mathbb{Z}/2, \mathbb{Z}/2)$ and $\mathrm{Ext}^{*,*}_{\Gamma(Sq^{0,1})}(\mathbb{Z}/2, \mathbb{Z}/2)$ are determined by the following equations:

$$Sq^k h_i = \begin{cases} h_{i+1} & \text{if } k = 0 \\ h_i^2 & \text{if } k = 1 \\ 0 & \text{if } k \geq 2 \end{cases} \tag{5.37}$$

$$Sq^k w_i = \begin{cases} w_{i+1} & \text{if } k = 0 \\ w_i^2 & \text{if } k = 1 \\ 0 & \text{if } k \geq 2 \end{cases} \tag{5.38}$$

and by the Cartan formula. The differential d_2 in the spectral sequence is determined by the formula

$$d_2(w_i) = h_i h_{i+1}, \qquad i \geq 0, \tag{5.39}$$

which is easily verified by considering the cobar construction on Λ. Then we can apply repeatedly the "Kudo Transgression Theorem" (Corollary 2.18 above) to conclude that for each $i \geq 0$ and $r \geq 0$, the class $w_i^{2^r}$ survives to E_{2^r+1}, and that

$$d_{2^r+1}[w_i^{2^r}]_{2^r+1} = [Sq^{2^{r-1}} Sq^{2^{r-2}} \dots Sq^1(h_i h_{i+1})]_{2^r+1}, \tag{5.40}$$

where the squaring operations on the right-hand side are to be computed using (5.37). For example, $d_3[w_i^2]_3 = [h_i^2 h_{i+2} + h_{i+1}^3]_3$. In interpreting Corollary 2.18 as in (5.40) we are of course using the fact that the squaring operations on the higher pages of the spectral sequence are determined by the operations on E_2 (Theorem 2.16), and that these in turn are given by (5.25a) and (5.25b).

But we wish to emphasize that the theory of operations in spectral seqauences that we have sketched in this work goes beyond the transgression theorem. So we consider a little further the spectral sequence for the extension (5.36). In $E_2^{1,1}$ we define classes θ_k for each $k \geq 0$ by

$$\theta_k = w_k h_{k+2} + w_{k+1} h_k. \tag{5.41}$$

Then we have from (5.39) that $d_2\theta_k = 0$. But in general the r'th differential maps $E_r^{p,q}$ to $E_r^{p+r,q-r+1}$; so, since $\theta_k \in E_2^{1,1}$ survives to E_3, it must be an infinite cycle. Then from Corollary 2.19,we have that the following elements of $E_2^{1,*}$:

$$(5.42) \qquad Sq^{2^{r-1}} Sq^{2^{r-2}} \cdots Sq^1\theta_k = w_k^{2^r} h_{k+r+2} + w_{k+1}^{2^r} h_{k+r} \quad (r \geq 1,\, k \geq 0)$$

are infinite cycles as well. The elements (5.41),(5.42), lying on the line p =1, can never be boundaries. Therefore, they represent a family of non-zero elements in $\mathrm{Ext}_\Lambda^{*,*}(\mathbb{Z}/2, \mathbb{Z}/2)$.

We remark finally that the vanishing of the higher differentials on the family (5.42) could not have been deduced solely from the geometry of the differentials in the $p - q$ plane. The Steenrod operations seem to play an essential role.

6. Application to the Cohomology of the Steenrod Algebra

The cohomology ring of the Steenrod algebra has been of great interest ever since Adams [**1**] identified it as the E_2 term of a spectral sequence converging to the stable homotopy groups of spheres. Here we describe some work of Palmieri [**64**] on the problem of determining this ring, and the application he makes of the theory developed here. We remark at the outset that the Hopf algebra extensions that Palmieri works with are not central, nor do the fibers have commutative multiplication. Thus, the results of neither [**86**] nor [**78**] would suffice. The theory developed in the present work, in its full generality, is required.

As in Wilkerson [**88**], a sub-Hopf algebra E of the Steenrod algebra $\mathcal{A}$ is called "elementary" if it is exterior as an algebra. Influenced by ideas introduced by Quillen into the cohomology theory of groups [**67**], Palmieri considers the set $\mathcal{Q}$ of elementary sub-Hopf algebras of $\mathcal{A}$ as a category: the morphisms are inclusions. As reviewed in Proposition 5.3, $\mathrm{Ext}_E^{*,*}(\mathbb{Z}/2, \mathbb{Z}/2)$ becomes a right algebra over $\overline{\mathcal{A}}$, the negative of the Steenrod algebra, when E is normal in $\mathcal{A}$. Using this fact Palmieri defines an action of $\overline{\mathcal{A}}$ on the inverse limit $\varprojlim_{\mathcal{Q}} \mathrm{Ext}_E^{*,*}(\mathbb{Z}/2, \mathbb{Z}/2)$. Restriction mappings in cohomology define a homomorphism of rings:

$$\gamma : \mathrm{Ext}_{\mathcal{A}}^{*,*}(\mathbb{Z}/2, \mathbb{Z}/2) \to (\varprojlim_{\mathcal{Q}} \mathrm{Ext}_E^{*,*}(\mathbb{Z}/2, \mathbb{Z}/2))^{\overline{\mathcal{A}}}.$$

Here the right hand side is the subring of the inverse limit consisting of those elements "invariant" under the $\overline{\mathcal{A}}$-action...i.e., those elements annihilated by the augmentation ideal. The main theorem of [**64**] asserts that γ is an F-isomorphism...that is, every element of the kernel is nilpotent; and that, if y is any element of the codomain, then y^{2^k} lies in the image of γ for some integer $k \geq 0$.

In order to prove the latter statement, Palmieri considers a diagram of Hopf algebra inclusions

$$\cdots \hookrightarrow D(n) \hookrightarrow D(n-1) \cdots \hookrightarrow D(0) = \mathcal{A}$$

where $D(n)$ is dual to a quotient of the dual Steenrod algebra:

$$(5.43) \qquad D(n) = \left(\mathbb{Z}/2[\xi_1, \xi_2, \xi_3 \ldots]/(\xi_1^2, \xi_2^4, \ldots, \xi_n^{2^n})\right)^*.$$

For each $n \geq 0$, $D(n)$ is normal is $\mathcal{A}$, and the action of $\overline{\mathcal{A}}$ on $\mathrm{Ext}_{D(n)}^{*,*}(\mathbb{Z}/2, \mathbb{Z}/2)$ passes to an action of the quotient $\overline{\mathcal{A}}//\overline{D}(n)$, as in Proposition 4.33. An important step in the proof that γ is an F-isomorphism is the following (Lemma 5.7 of [**64**]):

LEMMA 5.28. *For each* $y \in \left(\mathrm{Ext}^{*,*}_{D(n)}(\mathbb{Z}/2, \mathbb{Z}/2)\right)^{\overline{\mathcal{A}}//\overline{D}(n)}$ *there is an integer* j *such that* y^{2^j} *lifts to* $\left(\mathrm{Ext}^{*,*}_{D(n-1)}(\mathbb{Z}/2, \mathbb{Z}/2)\right)^{\overline{\mathcal{A}}//\overline{D}(n-1)}$.

Palmieri's proof of this result uses the change-of-rings spectral sequence for the extension:

$$D(n) \to D(n-1) \to D(n-1)//D(n). \tag{5.44}$$

For this spectral sequence one has:

$$E_2^{p,q} = \mathrm{Ext}^p_{D(n-1)//D(n)}(\mathbb{Z}/2, \overline{\mathrm{Ext}}^{D(n)}_{-q}(\mathbb{Z}/2, \mathbb{Z}/2)) \Longrightarrow \mathrm{Ext}^{p+q}_{D(n-1)}(\mathbb{Z}/2, \mathbb{Z}/2).$$

Any given $y \in \left(\mathrm{Ext}^{q,*}_{D(n)}(\mathbb{Z}/2, \mathbb{Z}/2)\right)^{\overline{\mathcal{A}}//\overline{D}(n)}$ is interpreted as an element of $E_2^{0,q}$. Using Theorem 2.21 repeatedly, Palmieri observes that for each positive integer i, y^{2^i} survives to $E^{0,*}_{2^i+1}$. He then chooses i so large that the only possible differentials:

$$d_j : E_j^{0,*,*} \to E_j^{j,l,t}, \qquad j \geq 2^i + 1 \tag{5.45}$$

land in those tri-degrees (j, l, t) for which he has established, by a separate "vanishing plane" argument, that $E_j^{j,l,t} = 0$. Here "t" is internal degree. Thus, for i sufficiently large, y^{2^i} survives to $E_\infty^{0,*,*}$, so the class $y^{2^i} \in \left(\mathrm{Ext}^{*,*}_{D(n)}(\mathbb{Z}/2, \mathbb{Z}/2)\right)^{\overline{\mathcal{A}}//\overline{D}(n)}$ lies in the image of the restriction mapping from $\left(\mathrm{Ext}^{*,*}_{D(n-1)}(\mathbb{Z}/2, \mathbb{Z}/2)\right)$. A separate argument (Lemma 4.4 of [**64**]) is necessary to show that some 2^kth power of y^{2^i} is actually the restriction of an invariant element; i.e., that for some $k \geq 0$, the element $y^{2^{i+k}}$ lies in $\left(\mathrm{Ext}^{*,*}_{D(n-1)}(\mathbb{Z}/2, \mathbb{Z}/2)\right)^{\overline{\mathcal{A}}//\overline{D}(n-1)}$.

Thus, the proof of Lemma 5.28 involves several steps. However, the theory of Steenrod operations in the change-of-rings spectral sequence, and Theorem 2.21 in particular, plays an important role.

7. Application to Finite sub-Hopf Algebras of the Steenrod Algebra

For each $n \geq 0$ write $\mathcal{A}(n)$ for the sub-Hopf algebra of the Steenrod algebra $\boldsymbol{\mathcal{A}}$ generated by the elements $\{Sq^1, Sq^2, Sq^4, \ldots, Sq^{2^n}\}$. The cohomology rings of the $\mathcal{A}(n)$ are known in the cases $n = 0, 1, 2$ ([**39**],[**75**]), but are difficult to describe in general. The methods described in this work seem to be a useful tool in the computation of these rings. We elaborate a little.

In his paper[**52**] Milnor describes the Hopf algebra dual to $\boldsymbol{\mathcal{A}}$ as a polynomial ring $\mathbb{Z}/2[\xi_1, \ldots, \xi_k, \ldots]$, with $|\xi_k| = 2^k - 1$, and coproduct determined by the formula:

$$\psi(\xi_k) = \sum_{i+j=k} \xi_i^{2^j} \otimes \xi_j.$$

For each $n \geq 1$ the ideal generated by the elements $\{\xi_1^{2^{n+1}}, \xi_2^{2^n}, \ldots, \xi_{n+1}^2, \xi_{n+2}, \ldots\}$ is a Hopf ideal, and the quotient Hopf algebra

$$\frac{\mathbb{Z}/2[\xi_1, \ldots, \xi_{n+1}, \xi_{n+2}, \ldots]}{(\xi_1^{2^{n+1}}, \xi_2^{2^n}, \ldots, \xi_{n+1}^2, \xi_{n+2}, \ldots)}$$

is dual to $\mathcal{A}(n)$ (see [**85**], pg. 23). We consider the following diagram of Hopf algebras:

$$\frac{\mathbb{Z}/2[\xi_1^2,\ldots,\xi_{n+1}^2]}{(\xi_1^{2^{n+1}},\xi_2^{2^n},\ldots,\xi_{n+1}^2)} \to \frac{\mathbb{Z}/2[\xi_1,\ldots,\xi_{n+1}]}{(\xi_1^{2^{n+1}},\xi_2^{2^n},\ldots,\xi_{n+1}^2)} \to E[\xi_1,\ldots,\xi_{n+1}]$$

where the end term is a primitively generated exterior algebra on the indicated elements. Dualizing this we obtain an extension of Hopf algebras:

$$E[Q_0,\ldots,Q_n] \to \mathcal{A}(n) \to D\mathcal{A}(n-1). \tag{5.46}$$

Here the fiber is a primitively generated exterior algebra, and Q_k is dual to ξ_{k+1}. $D\mathcal{A}(n-1)$ is the "double" of the Hopf algebra $\mathcal{A}(n-1)$; that is, the algebra and coalgebra structures are formally identical with those of $\mathcal{A}(n-1)$, but the gradings have been doubled. The cohomology of the fiber is a polynomial algebra on elements of homological degree 1:

$$\mathrm{Ext}^{*,*}_{E[Q_0,\ldots,Q_n]}(\mathbb{Z}/2,\mathbb{Z}/2) = \mathbb{Z}/2[v_0,v_1,\ldots,v_n]. \tag{5.47}$$

Here v_k is the element represented by ξ_{k+1} in the cobar construction. As reviewed in Proposition 4.33 the extension (5.46) defines a right action by the negative of $D\mathcal{A}(n-1)$ on the algebra (5.47), under which this polynomial ring becomes a right $D\overline{\mathcal{A}}(n-1)$ algebra. Straightforward computation gives for the action

$$(v_k)\cdot D(\overline{Sq}^{2^i}) = \begin{cases} v_{k-1} & \text{if } i=k-1 \\ 0 & \text{otherwise} \end{cases} \tag{5.48}$$

for all k with $1 \le k \le n$. One of the fundamental problems in computing the cohomology ring of $\boldsymbol{\mathcal{A}}(n)$ is to determine the image of the restriction mapping

$$\mathrm{Ext}^{*,*}_{\mathcal{A}(n)}(\mathbb{Z}/2,\mathbb{Z}/2) \to \left(\mathrm{Ext}^{*,*}_{E[Q_0,\ldots,Q_n]}(\mathbb{Z}/2,\mathbb{Z}/2)\right)^{D\overline{\mathcal{A}}(n-1)} \tag{5.49}$$

and in particular to determine which powers of the generators v_k lie in the image. It follows easily from (5.48) that for each $k \ge 1$, the lowest power of v_k invariant under action of $D\overline{\mathcal{A}}(n-1)$ is $v_k^{2^{n-k+1}}$ (this element being invariant since Sq^{2^n} is not present in $\mathcal{A}(n-1)$)! In [**44**], Conjecture 3.6, Mahowald and Shick conjecture that if $k \le n/2$ then precisely this power of v_k is present in the image of the restriction map; whereas if $k > n/2$ then only the powers of $v_k^{2^{k+1}}$ are present. They prove that $v_n^{2^{n+1}}$ is present in the cohomology of $\mathcal{A}(n)$, and claim that the truth of their conjecture in cases $k < n$ is supported by computation for small values of n.

The methods developed in the present work would seem effective tools in investigating questions of this kind. For the change-of-rings spectral sequence associated with the extension (5.46) one has:

(5.50)

$$E_2^{p,q,*} = \mathrm{Ext}^{p,*}_{D\mathcal{A}(n-1)}(\mathbb{Z}/2, \overline{\mathrm{Ext}}_{-q}^{E[Q_0,\ldots,Q_n]}(\mathbb{Z}/2,\mathbb{Z}/2)) \Longrightarrow \mathrm{Ext}^{p+q,*}_{\mathcal{A}(n)}(\mathbb{Z}/2,\mathbb{Z}/2).$$

In particular (we include internal degree):

$$E_2^{0,*,*} = \left(\mathrm{Ext}^{*,*}_{E[Q_0,\ldots,Q_n]}(\mathbb{Z}/2,\mathbb{Z}/2)\right)^{D\overline{\mathcal{A}}(n-1)}. \tag{5.51}$$

The elements of $E_2^{0,*,*}$ that survive to $E_\infty^{0,*,*}$ are the image of the restriction (5.49). This is the classical theorem that the edge homomorphism is the restriction mapping in cohomology. The vertical Steenrod operations act along this edge, and should be helpful in determining the survivors. Under the isomorphism (5.51) one sees easily that the vertical Steenrod operations described earlier in this chapter correspond to the Steenrod operations defined on the cohomology of the exterior Hopf algebra $E[Q_0, \ldots, Q_n]$, so that

$$Sq_V^{2^r}(v_k^{2^r}) = v_k^{2^{r+1}}$$

for all k with $0 \leq k \leq n$ and all $r \geq n-k+1$. Thus, knowledge of an early differential on a class $v_k^{2^{n-k+1}}$ will determine the value of later differentials on the higher powers of v_k, through the commutation rules of Steenrod squares with differentials developed in Chapter 2 of this work. Thus, the theory of Steenrod operations in the change-of-rings spectral sequence promises to be a useful tool in the study of the restriction mapping (5.49); and more generally in the study of the spectral sequence (5.50).

8. Applications to the Cohomology of Groups

Given any short-exact sequence of groups, one can pass to the associated sequence of group rings over $\mathbb{Z}/2$, obtaining an extension of Hopf algebras in the sense of Chapter 4. The change-of-rings spectral sequence for this extension is the Hochschild-Lyndon-Serre spectral sequence for the original extension of groups. The theory in the present work applies, and gives Steenrod operations in the spectral sequence.

The examples currently in the literature, so far as this writer knows, all involve the Kudo Transgression Theorem, Theorem 2.18. A form of this theorem adequate for the purpose is already in [**74**], Section II.9 c. For example, Benson points out in [**9**], pg. 120, that the cohomology rings of the dihedral and quaternionic groups of order eight can be computed using the transgression theorem. Quillen [**68**] has used the transgression theorem to calculate the cohomology rings of the "extra-special" 2-groups...groups that are extensions of an elementary abelian 2-group by a cyclic of order 2. A nice exposition of this work is given in [**11**].

It seems likely that the theory of Chapter 2 beyond the transgression theorem should be useful in the computation of the cohomology rings of groups. In particular, the commutation of Steenrod squares with differentials in general, as described in Theorem 2.17, should be of value.

CHAPTER 6

The Eilenberg-Moore Spectral Sequence

Let $\mathcal{G}$ be a simplicial group, and $\mathcal{E}$ a Kan complex which is also a principal right $\mathcal{G}$-space. Let $\mathcal{F}$ be a connected Kan complex on which $\mathcal{G}$ acts from the left. With homology and cohomology with coefficients in a fixed ground field l, we will construct the Eilenberg-Moore spectral sequence [**60**] in the form

$$E_2^{p,q} = \mathrm{Ext}^{p,q}_{H_*\mathcal{G}}(H_*\mathcal{E}, K_*\mathcal{F}) \Rightarrow H^{p+q}(\mathcal{E} \times_{\mathcal{G}} \mathcal{F}). \tag{6.1}$$

Here $K_*\mathcal{F}$ is the negatively graded cohomology of $\mathcal{F}$: $K_{-n}\mathcal{F} = H^n\mathcal{F}$ for all $n \in \mathbb{Z}$. In particular one could take $\mathcal{F}$ to be a point and $\mathcal{E}$ to be contractible, so that (6.1) becomes a spectral sequence for the cohomology of the classifying space of $\mathcal{G}$. In this chapter we will obtain (6.1) as the spectral sequence of a bisimplicial vector space over l. In the next chapter we will give the bisimplicial vector space the structure of a bisimplicial coalgebra, and specialize to the case $l = \mathbb{Z}/2$. Then the theory of Chapter 2 will apply, and automatically endow the spectral sequence with Steenrod operations.

We get the spectral sequence (6.1) by an argument different from the one originally used in [**60**]. The original argument used Serre's filtration of the chains of the fiber bundle $\mathcal{E}$, Serre's computation of the E_1-term of the resulting spectral sequence, and the somewhat technical "comparison theorem for spectral sequences" ([**59**], pgs. 4.17-4.22 and [**14**], pgs. 3.03-3.06). We do not use those results. Instead, we use some properties of semi-simplicial fibrations that were developed in [**7**], and that we review in Section 1 below.

The first main result of this chapter is Proposition 6.15. Here we construct a bisimplicial set $\mathcal{W}^{\mathcal{G}}(\mathcal{E}) \times_{\mathcal{G}} \mathcal{F}$ for which the associated diagonal simplicial set has the homotopy type of $\mathcal{E} \times_{\mathcal{G}} \mathcal{F}$. Then in Definition 6.17 we set up the Eilenberg-Moore spectral sequence, as the spectral sequence of the bisimplicial vector space spanned over l by $\mathcal{W}^{\mathcal{G}}(\mathcal{E}) \times_{\mathcal{G}} \mathcal{F}$. In Proposition 6.20 we compute the E_2 term of the spectral sequence.

We are writing in the semi-simplicial language, but the results are easily translated into the topological category. In fact, suppose $\underline{\mathcal{G}}$ is a topological group, $\underline{\mathcal{E}}$ a principal right $\underline{\mathcal{G}}$-bundle in the sense of [**84**], and $\underline{\mathcal{F}}$ a topological space on which $\underline{\mathcal{G}}$ acts from the left. Let Sing be the functor that to each topological space associates its singular simplicial set. Then Sing $\underline{\mathcal{G}}$ is a simplicial group, $\mathcal{G}$; Sing $\underline{\mathcal{E}}$ is a principal right $\mathcal{G}$-space, $\mathcal{E}$, and Sing $\underline{\mathcal{F}}$ a Kan complex $\mathcal{F}$ on which $\mathcal{G}$ acts from the left. Then the local product structure on $\underline{\mathcal{E}}$ can be used to show the simplicial sets $\mathcal{E} \times_{\mathcal{G}} \mathcal{F}$ and Sing$(\underline{\mathcal{E}} \times_{\underline{\mathcal{G}}} \underline{\mathcal{F}})$ are isomorphic, so that (6.1) is a spectral sequence for the cohomology of the Borel construction $\underline{\mathcal{E}} \times_{\underline{\mathcal{G}}} \underline{\mathcal{F}}$.

1. Kan Fibrations and Twisted Cartesian Products

In order to obtain (6.1) as the spectral sequence of a bisimplicial vector space we will need some standard results on Kan fibrations and twisted Cartesian products. These were originally developed in [**7**] and [**59**]. Another good reference is [**47**]. In this section we review the results we will need.

Let $\mathcal{G}$ be a simplicial group. By a "right $\mathcal{G}$-space" (left $\mathcal{G}$-space) we will mean a simplicial set on which $\mathcal{G}$ acts from the right (from the left). If $\mathcal{E}$ is a right $\mathcal{G}$-space, then $\mathcal{E} \times_{\mathcal{G}} \mathcal{PT}$ is the simplicial set consisting of the orbits of the action; here we are writing $\mathcal{PT}$ for the semi-simplicial point.

DEFINITION 6.1. Let $\mathcal{G}$ be a simplicial group, and $\mathcal{E}$ a right $\mathcal{G}$-space. We say $\mathcal{E}$ is a principal right $\mathcal{G}$-space if the equation $xg = x$ for some $g \in \mathcal{G}_n$ and some $x \in \mathcal{E}_n$ implies $g = e$, the identity element of $\mathcal{G}_n$. If this is the case, the projection mapping $\pi : \mathcal{E} \to \mathcal{E} \times_{\mathcal{G}} \mathcal{PT}$ is called a principal fibration with structural group $\mathcal{G}$ and base $\mathcal{B} = \mathcal{E} \times_{\mathcal{G}} \mathcal{PT}$.

DEFINITION 6.2. Let $\mathcal{B}$ be a simplicial set, and $\mathcal{G}$ a simplicial group. By a twisting function from $\mathcal{B}$ to $\mathcal{G}$ we mean a family of functions $\tau : \mathcal{B}_q \to \mathcal{G}_{q-1}$ defined for all $q > 0$, satisfying the following identities for all $b \in \mathcal{B}_q$:

$$\begin{array}{lcll}
d_0\tau(b) & = & (\tau d_0 b)^{-1} \cdot \tau d_1(b) & (q > 1) \\
d_i\tau(b) & = & \tau(d_{i+1}b) & (q > 1, i > 0) \\
s_i\tau(b) & = & \tau(s_{i+1}b) & (q > 0, i \geq 0) \\
\tau s_0(b) & = & e & (q \geq 0).
\end{array}$$

Now suppose in addition to the above data we are given a left $\mathcal{G}$-space $\mathcal{F}$. Define a simplicial set $\mathcal{B} \times_\tau \mathcal{F}$, called a twisted Cartesian product with fiber $\mathcal{F}$, base $\mathcal{B}$, and group $\mathcal{G}$, by writing:

$$(\mathcal{B} \times_\tau \mathcal{F})_n = \mathcal{B}_n \times \mathcal{F}_n.$$

Face and degeneracy operators are defined by:

$$\begin{array}{lcll}
d_0(b, f) & = & (d_0 b, \tau b \cdot d_0 f) & \\
d_i(b, f) & = & (d_i b, d_i f), & (i > 0) \\
s_i(b, f) & = & (s_i b, s_i f), & (i \geq 0).
\end{array}$$

That these operations satisfy the semi-simplicial identities is a consequence of the requirements on τ imposed above. The mapping $\pi : \mathcal{B} \times_\tau \mathcal{F} \to \mathcal{B}$ defined by $\pi(b, f) = b$ is clearly a mapping of simplicial sets.

PROPOSITION 6.3. *Let $\mathcal{B} \times_\tau \mathcal{F}$ be a twisted Cartesian product whose fiber $\mathcal{F}$ is a Kan complex. Then the mapping $\pi : \mathcal{B} \times_\tau \mathcal{F} \to B$ is a Kan fibration.*

This is proved in [**7**], [**59**], and in [**47**], Proposition 18.4.

A special case of this construction occurs if one takes $\mathcal{F} = \mathcal{G}$, with the natural left action of $\mathcal{G}$ upon itself. If a twisting function τ from $\mathcal{B}$ to $\mathcal{G}$ is given, the resulting simplicial set $\mathcal{B} \times_\tau \mathcal{G}$ is called a principal twisted Cartesian product.

Every principal twisted Cartesian product is a principal right $\mathcal{G}$-space: one defines the right action of $\mathcal{G}$ on $\mathcal{B} \times_\tau \mathcal{G}$ by writing $(b, g) \cdot g' = (b, gg')$. The converse is also true:

PROPOSITION 6.4. *Let $\pi : \mathcal{E} \to \mathcal{B}$ be a principal fibration with structural group $\mathcal{G}$ and base $\mathcal{B}$. Then there exists a twisting function τ from $\mathcal{B}$ to $\mathcal{G}$ and an isomorphism of simplicial sets $\mathcal{E} \to \mathcal{B} \times_\tau \mathcal{G}$ which commutes with right $\mathcal{G}$ action, and with the projections to $\mathcal{B}$.*

This is proved in [**7**],[**59**], and in [**47**], Proposition 18.7.

Now if $\pi : \mathcal{E} \to \mathcal{B}$ is a principal fibration with strucural group $\mathcal{G}$ and base $\mathcal{B}$, and if $\mathcal{F}$ is a left $\mathcal{G}$-space, we can form the "associated $\mathcal{G}$-bundle": the simplicial set $\mathcal{E} \times_\mathcal{G} \mathcal{F}$. Clearly the original projection $\pi : \mathcal{E} \to \mathcal{B}$ defines an associated projection $\pi : \mathcal{E} \times_\mathcal{G} \mathcal{F} \to \mathcal{B}$, by the formula $\pi(x, f) = \pi(x)$ for all $(x, f) \in \mathcal{E} \times_\mathcal{G} \mathcal{F}$. This can also be viewed as the mapping $\pi : \mathcal{E} \times_\mathcal{G} \mathcal{F} \to \mathcal{E} \times_\mathcal{G} \mathcal{PT}$ induced by the collapse $\mathcal{F} \to \mathcal{PT}$. As a consequence of the previous result we have:

PROPOSITION 6.5. *Let $\pi : \mathcal{E} \to \mathcal{B}$ be a principal fibration with structural group $\mathcal{G}$ and base $\mathcal{B}$. Let $\mathcal{F}$ be a left $\mathcal{G}$-space. Then the simplicial set $\mathcal{E} \times_\mathcal{G} \mathcal{F}$ can be identified with a twisted Cartesian product with fiber $\mathcal{F}$, base $\mathcal{B}$, and group $\mathcal{G}$ in a manner compatible with the projections to $\mathcal{B}$.*

PROOF. Let τ be a twisting function from $\mathcal{B}$ to $\mathcal{G}$ such that $\mathcal{E}$ can be identified with the principal twisted Cartesian product $\mathcal{B} \times_\tau \mathcal{G}$ in the manner described by Proposition 6.4. Then an easy computation shows that $\mathcal{E} \times_\mathcal{G} \mathcal{F}$ and $\mathcal{B} \times_\tau \mathcal{F}$ are isomorphic simplicial sets. □

COROLLARY 6.6. *Let $\pi : \mathcal{E} \to \mathcal{B}$ be a principal fibration with strucural group $\mathcal{G}$ and base $\mathcal{B}$. Let $\mathcal{F}$ be a left $\mathcal{G}$-space that is also a Kan complex. Then the mapping $\pi : \mathcal{E} \times_\mathcal{G} \mathcal{F} \to \mathcal{B}$ is a Kan fibration.*

This follows at once from Propositions 6.3 and 6.5.

We will need to know that certain simplicial sets appearing in Kan fibrations are Kan complexes. The following will be useful:

PROPOSITION 6.7. *Let $\pi : \mathcal{E} \to \mathcal{B}$ be a Kan fibration. Then:*
i. If $\mathcal{E}$ is a Kan complex and p is onto, then $\mathcal{B}$ is a Kan complex.
ii. If $\mathcal{B}$ is a Kan complex then $\mathcal{E}$ is a Kan complex.

This is [**47**], Proposition 7.5.

2. Bisimplicial Models for Fiber Bundles

Suppose given a simplicial group $\mathcal{G}$, a Kan complex $\mathcal{E}$ which is a principal right $\mathcal{G}$-space, and a Kan complex $\mathcal{F}$ on which $\mathcal{G}$ acts from the left; our goal in this section is to construct a certain bisimplicial set, which we call $\mathcal{W}^\mathcal{G}(\mathcal{E}) \times_\mathcal{G} \mathcal{F}$, for which the associated diagonal object has the homotopy type of $\mathcal{E} \times_\mathcal{G} \mathcal{F}$. The result is given in Proposition 6.15.

Our conventions for bisimplicial objects are reviewed in Chapter 1, Section 4. We begin with some elementary lemmas.

If H is any group, define a simplicial group $V(H)$ by setting $V_n(H) = H^{n+1}$. Here the face and degeneracy operators are given by the formulas $d_i(h_0, \ldots, h_n) = (h_0, \ldots, \widehat{h_i}, \ldots, h_n)$ and $s_i(h_0, \ldots, h_n) = (h_0, \ldots, h_i, h_i, \ldots, h_n)$. The group structure is given by coordinate-wise multiplication.

LEMMA 6.8. *If H is any group, the simplicial set $V(H)$ is contractible.*

PROOF. Since $V(H)$ is a simplicial group it is a Kan complex, so it suffices to show that $\pi_n(V(H)) = \{e\}$ for all $n \geq 0$. For each $n \geq 0$ define a morphism of sets $s : V_n(H) \to V_{n+1}(H)$ by setting $s(h_0, \ldots, h_n) = (e, h_0, \ldots, h_n)$, where $e \in H_{n+1}$ is the unit. One checks easily that $d_0 s(x) = x$, and that $d_1 s(x) = e$ if $x \in V_0(H)$ and that $d_{i+1} s(x) = s d_i(x)$ for all $i \geq 0$ if $x \in V_n(H)$ for $n > 0$. The existence of mappings s satisfying these equations implies the vanishing of all the homotopy groups of $V(H)$, and proves our Lemma. □

Now if $\mathcal{W}$ is a bisimplicial group, one can regard $\mathcal{W}_{*,q}$ for each fixed $q \geq 0$ as a simplicial group. One has:

LEMMA 6.9. *Let $\mathcal{W}$ be a bisimplicial group for which $\pi_p(\mathcal{W}_{*,q}) = \{e\}$ for all $p, q \geq 0$. Then* $\operatorname{Diag} \mathcal{W}$ *is contractible.*

PROOF. Quillen's spectral sequence of a bisimplicial group [**65**] implies that $\pi_* \operatorname{Diag} \mathcal{W} = e$ in all degrees. □

DEFINITION 6.10. Let $\mathcal{G}$ be a simplicial group. We define a functor T on the category of right $\mathcal{G}$-spaces by writing $T\mathcal{E} = \mathcal{E} \times \mathcal{G}$ for every right $\mathcal{G}$-space $\mathcal{E}$; the action of $\mathcal{G}$ being given by: $(x, g)g' = (xg', gg')$. For each right $\mathcal{G}$-space $\mathcal{E}$ define $d : T\mathcal{E} \to \mathcal{E}$ by writing $d(x, g) = x$ for all $x \in \mathcal{E}, g \in \mathcal{G}$. Then d is a natural transformation from T to the identity functor. For each right $\mathcal{G}$-space $\mathcal{E}$ define $s : T\mathcal{E} \to TT\mathcal{E}$ by writing $s(x, g) = (x, g, g)$ for all $x \in \mathcal{E}, g \in \mathcal{G}$. Then s is a natural transformation $s : T \to TT$. Clearly $\{T, d, s\}$ is a cotriple on the category of right $\mathcal{G}$-spaces.

As reviewed in Chapter 1, Section 11, this cotriple gives rise to a functor $\mathcal{W}^{\mathcal{G}}$ that to each right $\mathcal{G}$-space $\mathcal{E}$ associates a simplicial object $\mathcal{W}^{\mathcal{G}}(\mathcal{E})$ over the category of right $\mathcal{G}$-spaces. Thus,

$$\mathcal{W}^{\mathcal{G}}_p(\mathcal{E}) = T^{p+1}\mathcal{E} = \mathcal{E} \times \mathcal{G}^{p+1} \tag{6.2}$$

with face and degeneracy operators as described in Chapter 1. $\mathcal{W}^{\mathcal{G}}(\mathcal{E})$ comes with an augmentation

$$\lambda : \mathcal{W}^{\mathcal{G}}_0(\mathcal{E}) \to \mathcal{E} \tag{6.3}$$

which is just $d : T\mathcal{E} \to \mathcal{E}$.

Of course one can also regard $\mathcal{W}^{\mathcal{G}}(\mathcal{E})$ as a bisimplicial set, by the formula $\mathcal{W}^{\mathcal{G}}_{p,q}(\mathcal{E}) = (\mathcal{W}^{\mathcal{G}}_p(\mathcal{E}))_q$.

Given a simplicial group $\mathcal{G}$ write $\mathcal{PT}$ for the unique right $\mathcal{G}$-space consisting of a single point. Then $\mathcal{W}^{\mathcal{G}}_p(\mathcal{PT}) = \mathcal{G}^{(p+1)}$ for all $p \geq 0$. It is clear that $\mathcal{W}^{\mathcal{G}}(\mathcal{PT})$ becomes a bisimplicial group if one defines the multiplication coordinate-wise: $(g_0, \ldots, g_p) \cdot (h_0, \ldots, h_p) = (g_0 h_0, \ldots, g_p h_p)$.

PROPOSITION 6.11. *Let $\mathcal{G}$ be a simplicial group. Then the simplicial group* $\operatorname{Diag} \mathcal{W}^{\mathcal{G}}(\mathcal{PT})$ *is contractible.*

PROOF. By Lemma 6.9 it suffices to show that $\pi_p[\mathcal{W}^{\mathcal{G}}_{*,q}(\mathcal{PT})] = 0$ for all $p, q \geq 0$. But for each fixed q, the simplicial group $\mathcal{W}^{\mathcal{G}}_{*,q}(\mathcal{PT})$ is isomorphic to $V(\mathcal{G}_q)$, so our result follows from Lemma 6.8. □

Now we use $\operatorname{Diag} \mathcal{W}^{\mathcal{G}}(\mathcal{PT})$ to describe the structure of $\operatorname{Diag} \mathcal{W}^{\mathcal{G}}(\mathcal{E})$ for any right $\mathcal{G}$-space $\mathcal{E}$. In fact, if $\mathcal{G}$ is any simplicial group and $\mathcal{E}$ is a right $\mathcal{G}$-space one has $[\mathrm{Diag}\,\mathcal{W}^{\mathcal{G}}(\mathcal{E})]_p = \mathcal{E}_p \times (\mathcal{G}_p)^{p+1}$ with face and degeneracy operators given by:

$$\begin{aligned} d_i(x, g_0, \ldots, g_p) &= (d_i x, d_i g_0, \ldots, \widehat{d_i g_i}, \ldots, d_i g_p) \\ s_i(x, g_0, \ldots, g_p) &= (s_i x, s_i g_0, \ldots, s_i g_i, s_i g_i, \ldots, s_i g_p). \end{aligned}$$

From these formulae it is clear that $\mathrm{Diag}\,\mathcal{W}^{\mathcal{G}}(\mathcal{E})$ becomes a principal right $\mathcal{G}$-space if one defines the $\mathcal{G}$-action by: $(x, g_0, \ldots, g_p)\cdot g = (xg, g_0 g, \ldots, g_p g)$. It is also clear that we have an isomorphism of right $\mathcal{G}$-spaces:

$$\mathrm{Diag}\,\mathcal{W}^{\mathcal{G}}(\mathcal{E}) = \mathcal{E} \times \mathrm{Diag}\,\mathcal{W}^{\mathcal{G}}(\mathcal{PT}). \tag{6.4}$$

Since $\mathrm{Diag}\,\mathcal{W}^{\mathcal{G}}(\mathcal{PT})$ is a simplicial group, (6.4) implies that if $\mathcal{E}$ is a Kan complex then so is $\mathrm{Diag}\,\mathcal{W}^{\mathcal{G}}(\mathcal{E})$. Then from (6.4) and Proposition 6.11 we have:

PROPOSITION 6.12. *Let $\mathcal{G}$ be a simplicial group, and $\mathcal{E}$ a right $\mathcal{G}$-space that is also a Kan complex. The the mapping $\beta : \mathrm{Diag}\,\mathcal{W}^{\mathcal{G}}(\mathcal{E}) \to \mathcal{E}$ defined in each degree $p \geq 0$ by $\beta(x, g_0, \ldots, g_p) = x$ is a homotopy equivalence of Kan complexes.*

The mapping β of this proposition is a mapping of right $\mathcal{G}$-spaces. So we get an induced mapping of quotients: $\beta \times_{\mathcal{G}} \mathcal{PT} : (\mathrm{Diag}\,\mathcal{W}^{\mathcal{G}}(\mathcal{E})) \times_{\mathcal{G}} \mathcal{PT} \to \mathcal{E} \times_{\mathcal{G}} \mathcal{PT}$; and for any right $\mathcal{G}$-space $\mathcal{E}$ a commutative diagram:

$$\begin{array}{ccc} \mathrm{Diag}\,\mathcal{W}^{\mathcal{G}}(\mathcal{E}) & \xrightarrow{\ \beta\ } & \mathcal{E} \\ \downarrow & & \downarrow \\ (\mathrm{Diag}\,\mathcal{W}^{\mathcal{G}}(\mathcal{E})) \times_{\mathcal{G}} \mathcal{PT} & \xrightarrow[\beta \times_{\mathcal{G}} \mathcal{PT}]{} & \mathcal{E} \times_{\mathcal{G}} \mathcal{PT} \end{array} \tag{6.5}$$

Since $\mathrm{Diag}\,\mathcal{W}^{\mathcal{G}}(\mathcal{E})$ is always a principal right $\mathcal{G}$-space, we have from Corollary 6.6 (with $\mathcal{F} = \mathcal{G}$) that the left-hand vertical arrow in this diagram is a Kan fibration. If we assume in addition that $\mathcal{E}$ is a principal right $\mathcal{G}$-space then the right-hand vertical arrow is also a Kan fibration. If $\mathcal{E}$ is also a Kan complex then $\mathrm{Diag}\,\mathcal{W}^{\mathcal{G}}(\mathcal{E})$ is also Kan, and by Proposition 6.7, the base spaces in (6.5) are also Kan.

PROPOSITION 6.13. *Let $\mathcal{G}$ be a simplicial group, and $\mathcal{E}$ a principal right $\mathcal{G}$-space that is also a Kan complex. Then the mapping $\beta \times_{\mathcal{G}} \mathcal{PT}$ of (6.5) is a homotopy equivalence.*

PROOF. $\mathcal{E}$ can be written as a disjoint union of right $\mathcal{G}$-spaces $\mathcal{E}^i$ where for each index i, $\mathcal{G}_0$ acts transitively on the connected components of $\mathcal{E}^i$. Then by (6.4) we have also a decomposition of $\mathrm{Diag}\,\mathcal{W}^{\mathcal{G}}(\mathcal{E})$ as a disjoint union of the right $\mathcal{G}$-spaces $\mathrm{Diag}\,\mathcal{W}^{\mathcal{G}}(\mathcal{E}^i)$. Thus, it suffices to prove that for each i, $\beta \times_{\mathcal{G}} \mathcal{PT}$ is a homotopy equivalence from $(\mathrm{Diag}\,\mathcal{W}^{\mathcal{G}}(\mathcal{E}^i)) \times_{\mathcal{G}} \mathcal{PT}$ to $\mathcal{E}^i \times_{\mathcal{G}} \mathcal{PT}$. In other words, we may as well assume from the start that in (6.5), $\mathcal{G}_0$ acts transitively on the connected components of $\mathcal{E}$. Then by (6.4) and Proposition 6.11, $\mathcal{G}_0$ also acts transitively on the connected components of $\mathrm{Diag}\,\mathcal{W}^{\mathcal{G}}(\mathcal{E})$, so that both $\mathcal{E} \times_{\mathcal{G}} \mathcal{PT}$ and $(\mathrm{Diag}\,\mathcal{W}^{\mathcal{G}}(\mathcal{E})) \times_{\mathcal{G}} \mathcal{PT}$ are connected. Now (6.5) defines a mapping of the long-exact homotopy sequence of the Kan fibration on the left into the long exact homotopy sequence of that on the right. $\mathcal{G}$ is the common fiber; and by Proposition 6.12, β is induces isomorphisms of homotopy groups. So $\beta \times_{\mathcal{G}} \mathcal{PT}$ also induces isomorphisms of homotopy groups. But both $(\operatorname{Diag}\mathcal{W}^{\mathcal{G}}(\mathcal{E}))\times_{\mathcal{G}}\mathcal{PT}$ and $\mathcal{E}\times_{\mathcal{G}}\mathcal{PT}$ are connected Kan complexes, so $\beta\times_{\mathcal{G}}\mathcal{PT}$ is a homotopy equivalence. □

Now we wish to extend the above result, replacing the singleton space $\mathcal{PT}$ by more general class of left $\mathcal{G}$-spaces. So if $\mathcal{G}$ is a simplicial group, $\mathcal{E}$ a principal right $\mathcal{G}$-space, and $\mathcal{F}$ an arbitrary left $\mathcal{G}$-space, we consider the commutative diagram

(6.6)
$$\begin{array}{ccc}
(\operatorname{Diag}\mathcal{W}^{\mathcal{G}}(\mathcal{E}))\times_{\mathcal{G}}\mathcal{F} & \xrightarrow{\beta\times_{\mathcal{G}}\mathcal{F}} & \mathcal{E}\times_{\mathcal{G}}\mathcal{F} \\
\downarrow & & \downarrow \\
(\operatorname{Diag}\mathcal{W}^{\mathcal{G}}(\mathcal{E}))\times_{\mathcal{G}}\mathcal{PT} & \xrightarrow[\beta\times_{\mathcal{G}}\mathcal{PT}]{} & \mathcal{E}\times_{\mathcal{G}}\mathcal{PT}
\end{array}$$

where the vertical arrows are induced by the collapse $\mathcal{F}\to\mathcal{PT}$. By Proposition 6.5, each vertical arrow is the projection of a twisted Cartesian product to its base. If we assume $\mathcal{F}$ is a Kan complex then by Proposition 6.3 the vertical arrows are Kan fibrations. If in addition $\mathcal{E}$ is Kan we already know that both base spaces in (6.6) are Kan. So by Proposition 6.7, the total spaces in (6.6) are Kan as well.

PROPOSITION 6.14. *Let $\mathcal{G}$ be a simplicial group, and $\mathcal{E}$ a principal right $\mathcal{G}$-space that is also a Kan complex. Let $\mathcal{F}$ be a connected left $\mathcal{G}$-space that is also a Kan complex. Then the mapping $\beta\times_{\mathcal{G}}\mathcal{F}$ of* (6.6) *is a homotopy equivalence of Kan complexes.*

PROOF. By the same argument used in the proof of Proposition 6.13 we may assume without loss of generality that both base spaces in (6.6) are connected. Now (6.6) represents a mapping of Kan fibrations with common connected fiber $\mathcal{F}$. Thus the total spaces $(\operatorname{Diag}\mathcal{W}^{\mathcal{G}}(\mathcal{E}))\times_{\mathcal{G}}\mathcal{F}$ and $\mathcal{E}\times_{\mathcal{G}}\mathcal{F}$ are both connected, and it will suffice to show that $\beta\times_{\mathcal{G}}\mathcal{F}$ induces isomorphisms of homotopy groups. But from Proposition 6.13 we know that $\beta\times_{\mathcal{G}}\mathcal{PT}$ induces isomorphisms of homotopy, so our results follows from consideration of the mapping that (6.6) induces from the long-exact homotopy sequence of the fibration on the left to the long-exact homotopy sequence of the fibration on the right. □

We have promised the reader that we would construct a bisimplicial set for which the associated diagonal has the homotopy type of $\mathcal{E}\times_{\mathcal{G}}\mathcal{F}$. We can now do this easily. Define a simplicial object $\mathcal{W}^{\mathcal{G}}(\mathcal{E})\times_{\mathcal{G}}\mathcal{F}$ over the category of simplicial sets by writing: $(\mathcal{W}^{\mathcal{G}}(\mathcal{E})\times_{\mathcal{G}}\mathcal{F})_p=(\mathcal{W}^{\mathcal{G}}_p(\mathcal{E}))\times_{\mathcal{G}}\mathcal{F}$. Then $\mathcal{W}^{\mathcal{G}}(\mathcal{E})\times_{\mathcal{G}}\mathcal{F}$ can be regarded as a bisimplicial set, with:

(6.7)
$$(\mathcal{W}^{\mathcal{G}}(\mathcal{E})\times_{\mathcal{G}}\mathcal{F})_{p,q}=(\mathcal{W}^{\mathcal{G}}_p(\mathcal{E})\times_{\mathcal{G}}\mathcal{F})_q.$$

It is clear that:

$$\operatorname{Diag}(\mathcal{W}^{\mathcal{G}}(\mathcal{E})\times_{\mathcal{G}}\mathcal{F})=(\operatorname{Diag}\mathcal{W}^{\mathcal{G}}(\mathcal{E}))\times_{\mathcal{G}}\mathcal{F}.$$

So we have from Proposition 6.14:

PROPOSITION 6.15. *The mapping $\beta\times_{\mathcal{G}}\mathcal{F}$ of* (6.6) *is a homotopy equivalence:*

(6.8)
$$\beta\times_{\mathcal{G}}\mathcal{F}:\operatorname{Diag}(\mathcal{W}^{\mathcal{G}}(\mathcal{E})\times_{\mathcal{G}}\mathcal{F})\to\mathcal{E}\times_{\mathcal{G}}\mathcal{F}.$$

Finally we note the augmentation (6.3), being a mapping of right $\mathcal{G}$-spaces, extends to an augmentation:

$$\lambda : (\mathcal{W}_0^{\mathcal{G}}(\mathcal{E}) \times_{\mathcal{G}} \mathcal{F}) \to \mathcal{E} \times_{\mathcal{G}} \mathcal{F}. \tag{6.9}$$

This mapping of simplicial sets is just the usual projection $\lambda : \mathcal{E} \times \mathcal{F} \to \mathcal{E} \times_{\mathcal{G}} \mathcal{F}$.

3. Construction of the Spectral Sequence

In the next two sections we fix a simplicial group $\mathcal{G}$, a principal right $\mathcal{G}$-space $\mathcal{E}$ and a connected left $\mathcal{G}$-space $\mathcal{F}$; we assume both $\mathcal{E}$ and $\mathcal{F}$ are Kan. We fix a ground field l. Our goal in this section is to define a certain bisimplicial vector space $W^G(E) \times_G F$, for which the associated total complex has cohomology equal to that of $\mathcal{E} \times_{\mathcal{G}} \mathcal{F}$. This result is given in Proposition 6.16. Then we will define the Eilenberg-Moore spectral sequence (6.1) to be the spectral sequence of the bisimplicial vector space $W^G(E) \times_G F$. This is done in Definition 6.17.

We write E, F, G, and $W_p^G(E)$ for the simplicial vector spaces with bases $\mathcal{E}$, $\mathcal{F}$, $\mathcal{G}$ and $\mathcal{W}_p^{\mathcal{G}}(\mathcal{E})$ respectively. If $\mathcal{W}$ is a right $\mathcal{G}$-space and $\mathcal{F}$ a left $\mathcal{G}$-space we will write $W \times_G F$ for the simplicial vector space with basis $\mathcal{W} \times_{\mathcal{G}} \mathcal{F}$. Thus, formally speaking, in each dimension n we are setting

$$(W \times_G F)_n = W_n \otimes_{l(\mathcal{G}_n)} F_n$$

where $l(\mathcal{G}_n)$ is the group ring of $\mathcal{G}_n$.

We adopt similar notations for bisimplicial objects. We will write $W^G(E)$ and $W^G(E) \times_G F$ for the bisimplicial vector spaces with bases $\mathcal{W}^{\mathcal{G}}(\mathcal{E})$ and $\mathcal{W}^{\mathcal{G}}(\mathcal{E}) \times_G \mathcal{F}$, respectively. As in (6.7) we have:

$$(W^G(E) \times_G F)_{p,q} = (W_p^G(E) \times_G F)_q.$$

Associated with these bisimplicial vector spaces are the corresponding "total" chain complexes $CW^G(E), C(W^G(E) \times_G F)$, as reviewed in Chapter 1, Section 15. The augmentations (6.3), (6.9) pass to augmentations of bisimplicial vector spaces, in the sense of Definition 1.123: $\lambda : W_{0,*}^G(E) \to E_*$ and $\lambda : (W^G(E) \times_G F)_{0,*} \to (E \times_G F)_*$. As in (1.135) we have associated chain mappings: $C\lambda : CW^G(E) \to CE$ and $C\lambda : C(W^G(E) \times_G F) \to C(E \times_G F)$. We claim the following diagrams of chain mappings commute:

$$\begin{array}{ccc} C \operatorname{Diag} W^G(E) & \xrightarrow{C\beta} & CE \\ {\scriptstyle f}\downarrow & \nearrow {\scriptstyle C\lambda} & \\ CW^G(E) & & \end{array} \tag{6.10}$$

and

$$\begin{array}{ccc} C \operatorname{Diag}(W^G(E) \times_G F) & \xrightarrow{C(\beta \times_G \mathcal{F})} & C(E \times_G F) \\ {\scriptstyle f}\downarrow & \nearrow {\scriptstyle C\lambda} & \\ C(W^G(E) \times_G F) & & \end{array} \tag{6.11}$$

Here f is the chain equivalence of Eilenberg and Mac Lane as given by (1.136), and the chain mappings $C\beta$ and $C(\beta \times_G \mathcal{F})$ are defined by β of Proposition 6.12 in the obvious way. In fact commutativity of (6.10) is established by direct computation at the chain level, and the commutativity of (6.11) is an easy consequence. Then from Propositions 1.124 and 6.15 we have:

PROPOSITION 6.16. *The augmentation* $\lambda : (W^G(E) \times_G F)_{0,*} \to (E \times_G F)_*$ *induces an isomorphism of cohomology groups:*

$$\lambda^* : H^*(\mathcal{E} \times_{\mathcal{G}} \mathcal{F}) \to H^* \operatorname{Hom}_l(C(W^G(E) \times_G F), l).$$

DEFINITION 6.17. Suppose given a simplicial group $\mathcal{G}$, a principal right $\mathcal{G}$-space $\mathcal{E}$ and a connected simplicial set $\mathcal{F}$ on which $\mathcal{G}$ acts from the left. Assume $\mathcal{E}$ and $\mathcal{F}$ both Kan. By the associated Eilenberg-Moore spectral sequence we mean the spectral sequence of the bisimplicial l-module $W^G(E) \times_G F$, with coefficients in l, as discussed in Chapter 1, Section 16.

The spectral sequence converges to $H^* \operatorname{Hom}_l(C(W^G(E) \times_G F), l)$, and so, by Proposition 6.16, to $H^*(\mathcal{E} \times_{\mathcal{G}} \mathcal{F})$.

4. Calculation of the E_2-Term

In this section we compute the E_2-term of the Eilenberg-Moore spectral sequence. We will use the material on differential algebras and differential modules reviewed in Chapter 1, Sections 8 and 10.

We continue to use the convention of the previous section: if $\mathcal{R}$ is an arbitrary simplicial set we write R for the simplicial l-vector space that is spans.

Let T be the functor on the category of right $\mathcal{G}$-spaces described in Definition 6.10.

LEMMA 6.18. *Let* $\mathcal{E}$ *be any right* $\mathcal{G}$*-space. Then* $T\mathcal{E}$ *is isomorphic to an extended right* $\mathcal{G}$*-space.*

PROOF. If $\mathcal{E}$ is a given right $\mathcal{G}$-space we define a mapping of simplicial sets $\phi : \mathcal{E} \times \mathcal{G} \to \mathcal{E} \times \mathcal{G}$ by setting $\phi(x, g) = (xg^{-1}, g)$ for all $x \in \mathcal{E}$, $g \in \mathcal{G}$. If the domain of ϕ is interpreted as the $\mathcal{G}$-space $T\mathcal{E}$ and the codomain as the extended $\mathcal{G}$-space as in Definition 1.106, then it is easily seen that ϕ is a morphism of right $\mathcal{G}$-spaces, and in fact is an isomorphism. □

As a consequence of Lemmas 1.107 and 6.18 we have:

COROLLARY 6.19. *Let* $\mathcal{E}$ *be a right* $\mathcal{G}$*-space and* $\mathcal{F}$ *a left* $\mathcal{G}$*-space. Then the mapping* (1.97) *gives an isomorphism of graded vector spaces:*

$$H^*(T\mathcal{E} \times_{\mathcal{G}} \mathcal{F}) \to \operatorname{Hom}_{H_*\mathcal{G}}(H_*T\mathcal{E}, K_*\mathcal{F}).$$

Now we can describe the E_2-term of the Eilenberg-Moore spectral sequence. We start with (1.149) and Definition 6.17, according to which, for each fixed $q \geq 0$, $E_2^{*,q}$ is the homology of the complex

$$H^q(\mathcal{W}_0^{\mathcal{G}}(\mathcal{E}) \times_{\mathcal{G}} \mathcal{F}) \to \cdots \to H^q(\mathcal{W}_{p-1}^{\mathcal{G}}(\mathcal{E}) \times_{\mathcal{G}} \mathcal{F}) \xrightarrow{\delta_h} H^q(\mathcal{W}_p^{\mathcal{G}}(\mathcal{E}) \times_{\mathcal{G}} \mathcal{F}) \to \cdots .$$

Here the "horizontal" differential δ_h is the alternating sum of coface operators δ_h^i, each of which is the mapping in cohomology induced by the i'th face map $d_i : \mathcal{W}_p^{\mathcal{G}}(\mathcal{E}) \to \mathcal{W}_{p-1}^{\mathcal{G}}(\mathcal{E})$. Since the right $\mathcal{G}$-space $\mathcal{W}_p^{\mathcal{G}}(\mathcal{E})$ is in the image of the functor T for each $p \geq 0$ we can use Corollary 6.19 and the naturality of (1.97) to rewrite this complex in the form:

$$(6.12)\qquad \mathrm{Hom}^q_{H_*\mathcal{G}}(H_*\mathcal{W}^{\mathcal{G}}_0(\mathcal{E}), K_*\mathcal{F}) \to \cdots \to \mathrm{Hom}^q_{H_*\mathcal{G}}(H_*\mathcal{W}^{\mathcal{G}}_{p-1}(\mathcal{E}), K_*\mathcal{F}) \text{——} \xrightarrow{\delta_h} \mathrm{Hom}^q_{H_*\mathcal{G}}(H_*\mathcal{W}^{\mathcal{G}}_p(\mathcal{E}), K_*\mathcal{F}) \to \cdots .$$

In this case, δ_h is the alternating sum of coface operators δ^i_h, each of which is determined by the mapping $H_*d_i : H_*\mathcal{W}^{\mathcal{G}}_p(\mathcal{E}) \to H_*\mathcal{W}^{\mathcal{G}}_{p-1}(\mathcal{E})$ of right $H_*\mathcal{G}$-modules.

It is easy to interpret the homology groups of the complex (6.12). Recall that, associated to the Hopf algebra $H_*\mathcal{G}$ is a canonical cotriple $\{T, d, s\}$ on the category of right $H_*\mathcal{G}$-modules, as reviewed in Chapter 1, Example 1.108. Abusing notation a little, we will denote by $\{T, d, s\}$ both this cotriple, and the cotriple on the category of right $\mathcal{G}$-spaces described in Definition 6.10. These cotriples are closely related. The Eilenberg-Zilber and Künneth theorems give a natural equivalence $TH_* \to H_*T$, and the following square commutes:

$$\begin{array}{ccccc}
TH_*\mathcal{E} & = & H_*\mathcal{E} \otimes H_*\mathcal{G} & \xrightarrow{d} & H_*\mathcal{E} \\
\downarrow & & \downarrow & & \downarrow Id \\
H_*T\mathcal{E} & = & H_*(\mathcal{E} \times \mathcal{G}) & \xrightarrow[H_*d]{} & H_*\mathcal{E}
\end{array}$$

It follows that for each $p \geq 0$ and for each i with $0 \leq i \leq p$ we get a commutative diagram of right modules over $H_*\mathcal{G}$:

$$(6.13)\qquad \begin{array}{ccc}
W^{H_*\mathcal{G}}_p(H_*\mathcal{E}) & \xrightarrow{d_i} & W^{H_*\mathcal{G}}_{p-1}(H_*\mathcal{E}) \\
\downarrow & & \downarrow \\
H_*\mathcal{W}^{\mathcal{G}}_p(\mathcal{E}) & \xrightarrow[H_*d_i]{} & H_*\mathcal{W}^{\mathcal{G}}_{p-1}(\mathcal{E})
\end{array}$$

Here the d_i in the top row is a face map of the simplicial object $W^{H_*\mathcal{G}}(H_*\mathcal{E})$, as described in Example 1.108. The d_i in the bottom row is a face map of $\mathcal{W}^{\mathcal{G}}(\mathcal{E})$, and the vertical arrows are isomorphisms coming from the Künneth theorem. So we can rewrite (6.12): $E_2^{p,q}$ is the p'th homology group of the complex

$$\cdots \longrightarrow \mathrm{Hom}^q_{H_*\mathcal{G}}(W^{H_*\mathcal{G}}_{p-1}(H_*\mathcal{E}), K_*\mathcal{F}) \xrightarrow{\delta_h} \mathrm{Hom}^q_{H_*\mathcal{G}}(W^{H_*\mathcal{G}}_p(H_*\mathcal{E}), K_*\mathcal{F}) \longrightarrow \cdots$$

where δ_h is the dual of the differential $\partial = \sum_{i=0}^p (-1)^i d_i$ in the chain complex $CW^{H_*\mathcal{G}}(H_*\mathcal{E})$. But as reviewed in Example 1.108, $CW^{H_*\mathcal{G}}(H_*\mathcal{E})$ is a free resolution of $H_*\mathcal{E}$ as an $H_*\mathcal{G}$-module. So we have at once:

PROPOSITION 6.20. *The E_2-term of the Eilenberg-Moore spectral sequence is given by:*

$$E_2^{p,q} = \mathrm{Ext}^{p,q}_{H_*\mathcal{G}}(H_*\mathcal{E}, K_*\mathcal{F}). \tag{6.14}$$

This completes our construction of the Eilenberg-Moore spectral sequence (6.1).

CHAPTER 7

Steenrod Operations in the Eilenberg-Moore Spectral Sequence

We continue to assume $\mathcal{G}$ a simplicial group, $\mathcal{E}$ a principal right $\mathcal{G}$-space that is also a Kan complex, and $\mathcal{F}$ a Kan complex on which $\mathcal{G}$ acts from the left. We will use the theory developed in Chapter 2 to put Steenrod operations into the Eilenberg-Moore spectral sequence (6.1), and also to describe the operations at the E_2-level. In order to accomplish this latter task we will need to assume that the graded vector spaces $H_*\mathcal{G}$, $H_*\mathcal{E}$ and $H_*\mathcal{F}$ are of finite type, and will do so without further comment throughout this chapter.

The results of this chapter were originally sketched in [**78**]. They have also been worked out, from a somewhat different point of view, by Mori in [**61, 62**], where the odd-prime case is treated as well.

In Section 1 we derive the basic results on the action of the Steenrod operations in the spectral sequence, and show the operations compatible with those defined on the cohomology of the target $H^*(\mathcal{E} \times_{\mathcal{G}} \mathcal{F})$. In Section 2 we use the language of homological algebra to define vertical and diagonal Steenrod operations on $\mathrm{Ext}^{*,*}_{H_*\mathcal{G}}(H_*\mathcal{E}, K_*\mathcal{F})$. In Section 3 we show that these algebraically defined operations agree with the operations on the E_2-term of the spectral sequence, as these were computed in Chapter 2. In Section 4 we briefly describe some of the applications that have been made of these results in the computation of the cohomology of classifying spaces.

1. The Spectral Sequence with its Products and Steenrod Squares

We begin by observing that since $E \times_G F$ is a simplicial vector space with basis a simplicial set, it becomes a simplicial coalgebra with counit and coproduct determined by (1.73). For each $p \geq 0$, note that $(W^G(E) \times_G F)_{p,*}$ is the simplicial vector space with basis the simplicial set $(W^{\mathcal{G}}(\mathcal{E}) \times_{\mathcal{G}} \mathcal{F})_{p,*}$, so it also becomes a simplicial coalgebra under the definitions (1.73). The horizontal simplicial operators in $W^G(E) \times_G F$ are coalgebra morphisms, since they come from mappings of simplicial sets. So $W^G(E) \times_G F$ is a bisimplicial coalgebra over the ground field l. Clearly, the augmentation $\lambda : (W^G(E) \times_G F)_{0,*} \to (E \times_G F)_*$ is a morphism of simplicial coalgebras.

Now we take $l = \mathbb{Z}/2$, and we write Hom for $\mathrm{Hom}_{\mathbb{Z}/2}$. In this case both $H^*\,\mathrm{Hom}(C(E \times_G F), \mathbb{Z}/2)$ and $H^*\,\mathrm{Hom}(C(W^G(E) \times_G F), \mathbb{Z}/2)$ admit products and Steenrod operations ((1.121), (1.122), (2.12), (2.17)). The products and Steenrod operations defined on $H^*(\mathrm{Hom}(C(E \times_G F), \mathbb{Z}/2))$ are those classically defined on the cohomology of the space $\mathcal{E} \times_{\mathcal{G}} \mathcal{F}$. Sq^0 is the identity operation in such a case; and so, as we observed in Chapter 1, Section 13, the Steenrod squares acting on

$H^*(\mathcal{E} \times_{\mathcal{G}} \mathcal{F}, \mathbb{Z}/2)$ define an action of the Steenrod algebra $\boldsymbol{\mathcal{A}}$. From Propositions 2.1 and 6.16 we have:

PROPOSITION 7.1. *The mapping*

$$\lambda^* : H^*(\mathcal{E} \times_{\mathcal{G}} \mathcal{F}; \mathbb{Z}/2) \to H^* \operatorname{Hom}(C(W^G(E) \times_G F), \mathbb{Z}/2)$$

is an isomorphism of $\boldsymbol{\mathcal{A}}$-algebras.

The effect of this proposition is to assure that when, in the next theorem, we put products and squaring operations into the spectral sequence of the bisimplicial coalgebra $W^G(E) \times_G F$, we will know that these converge to the operations ordinarily defined on the cohomology of $\mathcal{E} \times_{\mathcal{G}} \mathcal{F}$.

We apply the theory of Chapter 2 to the spectral sequence of the bisimplicial coalgebra $W^G(E) \times_G F$ with coefficients in $\mathbb{Z}/2$. The role of the bialgebra Λ in Chapter 2 is here played simply by $\mathbb{Z}/2$. We have at once:

THEOREM 7.2. *Let $\mathcal{G}$ be a simplicial group, and $\mathcal{E}$ a principal right $\mathcal{G}$-space that is also a Kan complex. Let $\mathcal{F}$ be a Kan complex on which $\mathcal{G}$ acts from the left. Then the Eilenberg-Moore spectral sequence for $H^*(\mathcal{E} \times_{\mathcal{G}} \mathcal{F}; \mathbb{Z}/2)$ admits products*

$$\mu : E_r^{*,*} \otimes E_r^{*,*} \to E_r^{*,*}$$

and squaring operations

$$\begin{aligned} Sq^k &: E_r^{p,q} \to E_r^{p,q+k} && \text{if } 0 \le k \le q \\ Sq^k &: E_r^{p,q} \to E_{r+k-q}^{p+k-q,2q} && \text{if } q \le k \le q+r-2 \\ Sq^k &: E_r^{p,q} \to E_{2r-2}^{p+k-q,2q} && \text{if } q+r-2 \le k \end{aligned}$$

for all $r \geq 2$, satisfying all propositions and theorems of Chapter 2, Sections 2 and 3. In particular, the operations on E_r are determined by the operations at the E_2-level as described in Theorem 2.16. Products and squaring operations commute with the differentials of the spectral sequence as described in Proposition 2.10 and Theorem 2.17. Products and squaring operations are defined on E_∞ and are compatible with the operations on the target of the spectral sequence in the manner described by Proposition 2.11 and Theorem 2.22. If $\alpha \in E_r$ then α^2 survives to E_{2r-1}, as described in Theorem 2.21.

2. Steenrod Operations on $\operatorname{Ext}^{*,*}_{H_*\mathcal{G}}(H_*\mathcal{E}, K_*\mathcal{F})$

In this section we define, purely in the language of homological algebra, a product:

$$\mu : \operatorname{Ext}^{*,*}_{H_*\mathcal{G}}(H_*\mathcal{E}, K_*\mathcal{F}) \otimes \operatorname{Ext}^{*,*}_{H_*\mathcal{G}}(H_*\mathcal{E}, K_*\mathcal{F}) \to \operatorname{Ext}^{*,*}_{H_*\mathcal{G}}(H_*\mathcal{E}, K_*\mathcal{F}) \tag{7.1}$$

and squaring operations:

$$Sq^k_V : \operatorname{Ext}^{p,q}_{H_*\mathcal{G}}(H_*\mathcal{E}, K_*\mathcal{F}) \to \operatorname{Ext}^{p,q+k}_{H_*\mathcal{G}}(H_*\mathcal{E}, K_*\mathcal{F}) \tag{7.2a}$$

$$Sq^k_D : \operatorname{Ext}^{p,q}_{H_*\mathcal{G}}(H_*\mathcal{E}, K_*\mathcal{F}) \to \operatorname{Ext}^{p+k,2q}_{H_*\mathcal{G}}(H_*\mathcal{E}, K_*\mathcal{F}). \tag{7.2b}$$

In the next section we will identify these operations with the product and with the vertical and diagonal Steenrod squares on the E_2-term of the Eilenberg-Moore spectral sequence, as these were defined in Chapter 2.

We will use the theory of Chapter 1, Section 14 to define the operations (7.1) and (7.2b). We start by giving $H_*\mathcal{G}$ its classical structure as a Hopf algebra as reviewed in Chapter 1, Section10. We must also define structures of an $H_*\mathcal{G}$-coalgebra for $H_*\mathcal{E}$, and an $H_*\mathcal{G}$-algebra for $K_*\mathcal{F}$. The first of these is already explicit in Proposition 1.102, which asserts that the coproduct $\psi_{\mathcal{E}} : H_*\mathcal{E} \to H_*\mathcal{E} \otimes H_*\mathcal{E}$ is linear over $H_*\mathcal{G}$. Similarly, we give $H_*\mathcal{F}$ the structure of a left $H_*\mathcal{G}$-coalgebra. Then the negatively graded dual $K_*\mathcal{F} = \overline{\mathrm{Hom}}(H_*\mathcal{F}, \mathbb{Z}/2)$ acquires the dual, right action of $H_*\mathcal{G}$ as in (1.25). Defining the product on $K_*\mathcal{F}$ as in (1.31), dual to the coproduct on $H_*\mathcal{F}$, we have from Proposition 1.31 that $K_*\mathcal{F}$ is a right $H_*\mathcal{G}$-algebra.

DEFINITION 7.3. By the product (7.1) and diagonal Steenrod squares (7.2b) we mean the product and Steenrod squares that are ordinarily defined on the cohomology of the Hopf algebra $H_*\mathcal{G}$ with coefficients in the $H_*\mathcal{G}$-coalgebra $H_*\mathcal{E}$ and commutative $H_*\mathcal{G}$-algebra $K_*\mathcal{F}$, as in (1.125) and (1.126).

As an immediate consequence of Proposition 1.120 our diagonal squares give $\mathrm{Ext}^{*;*}_{H_*\mathcal{G}}(H_*\mathcal{E}, K_*\mathcal{F})$ the structure of a left $\boldsymbol{\mathcal{H}}$-algebra:

PROPOSITION 7.4. *The diagonal Steenrod squares* (7.2b) *satisfy the Adem relations, and satisfy the Cartan formula with respect to the product* (7.1).

To define the vertical squares (7.2a) we are going to use the theory developed in Chapter 3, Section 1, with the substitutions:

$$\Pi \to \overline{\boldsymbol{\mathcal{A}}}, \quad \Xi \to H_*\mathcal{G}, \quad M \to H_*\mathcal{E}, \quad N \to K_*\mathcal{F}. \tag{7.3}$$

In particular we want to employ Proposition 3.14 and Definition 3.20, and so must first define:

1. on $H_*\mathcal{G}$, the structure of a right $\overline{\boldsymbol{\mathcal{A}}}$-algebra;
2. on $H_*\mathcal{E}$, the structure of a right $\overline{\boldsymbol{\mathcal{A}}} - H_*\mathcal{G}$ module;
3. on $K_*\mathcal{F}$, the structure of a module with compatible actions of $\overline{\boldsymbol{\mathcal{A}}}$ on the left and $H_*\mathcal{G}$ on the right, in the sense of Definition 3.8.

We recall first that since we are assuming $H_*\mathcal{G}, H_*\mathcal{E}$ and $H_*\mathcal{F}$ all of finite type, they acquire right actions of $\overline{\boldsymbol{\mathcal{A}}}$, dual to the left actions of $\boldsymbol{\mathcal{A}}$ on $H^*\mathcal{G}, H^*\mathcal{E}$ and $H^*\mathcal{F}$, as in Remark 1.116. According to Corollary 1.119 these actions make $H_*\mathcal{G}$ into a right $\overline{\boldsymbol{\mathcal{A}}}$-algebra and $H_*\mathcal{E}$ into a right $\overline{\boldsymbol{\mathcal{A}}} - H_*\mathcal{G}$ module, so we have defined the first two structures listed above. To get the third, we use Example 3.9. Regard $H_*\mathcal{F}$ as a right $\overline{\boldsymbol{\mathcal{A}}} - H_*\mathcal{G}$ module, with left action of $H_*\mathcal{G}$. Then $K_*\mathcal{F} = \overline{\mathrm{Hom}}(H_*\mathcal{F}, \mathbb{Z}/2)$ acquires the dual actions of $\overline{\boldsymbol{\mathcal{A}}}$ from the left and $H_*\mathcal{G}$ from the right. As observed in Example 3.9, these actions are "compatible". So we have the third structure listed above. It follows that the vector space $\mathrm{Hom}_{H_*\mathcal{G}}(H_*\mathcal{E}, K_*\mathcal{F})$ becomes a left $\boldsymbol{\mathcal{A}}$-module, with action defined as in (3.1) and Proposition 3.14:

$$(Sq^k \cdot f)(m) = \sum_{i+j=k} \overline{Sq^i} f(m\overline{Sq^j}) \tag{7.4}$$

for each $f \in \mathrm{Hom}_{H_*\mathcal{G}}(H_*\mathcal{E}, K_*\mathcal{F})$, $m \in H_*\mathcal{E}$, $k \geq 0$.

We follow Proposition 3.14 and Definition 3.20. Let $\mathcal{P}(H_*\mathcal{E}) \to H_*\mathcal{E} \to 0$ be any resolution of $H_*\mathcal{E}$ by relative projectives in the category of $\overline{\boldsymbol{\mathcal{A}}} - H_*\mathcal{G}$ modules. For each $p \geq 0$, the vector space $\mathrm{Hom}_{H_*\mathcal{G}}(\mathcal{P}_p(H_*\mathcal{E}), K_*\mathcal{F})$ becomes a left $\boldsymbol{\mathcal{A}}$-module, with action defined by (7.4) for each $f \in \mathrm{Hom}_{H_*\mathcal{G}}(\mathcal{P}_p(H_*\mathcal{E}), K_*\mathcal{F})$, and for all $m \in \mathcal{P}_p(H_*\mathcal{E})$, $k \geq 0$. From the naturality clause in Proposition 3.14, one sees that $\mathrm{Hom}_{H_*\mathcal{G}}(\mathcal{P}(H_*\mathcal{E}), K_*\mathcal{F})$ becomes a cochain complex over the category of left $\boldsymbol{\mathcal{A}}$-modules; so that, as in Definition 3.20, the equation

$$\mathrm{Ext}^{p,*}_{H_*\mathcal{G}}(H_*\mathcal{E}, K_*\mathcal{F}) = H^{p,*}\,\mathrm{Hom}_{H_*\mathcal{G}}(\mathcal{P}(H_*\mathcal{E}), K_*\mathcal{F}) \tag{7.5}$$

defines $\mathrm{Ext}^{p,*}_{H_*\mathcal{G}}(H_*\mathcal{E}, K_*\mathcal{F})$ as a left $\boldsymbol{\mathcal{A}}$-module for each $p \geq 0$.

DEFINITION 7.5. By the vertical squares (7.2a) defined on the E_2-term of the Eilenberg-Moore spectral sequence, we mean those associated with the $\boldsymbol{\mathcal{A}}$-action on $\mathrm{Ext}^{p,*}_{H_*\mathcal{G}}(H_*\mathcal{E}, K_*\mathcal{F})$ as given by (7.4) and (7.5).

We would next like to study the relationship between the vertical squares, on one hand, and the product and diagonal squares on the other. Our aim is to invoke Proposition 3.21, so we must check that the hypotheses of that proposition are satisfied in the present context. In addition to items 1-3 listed above, we need:

4. on $H_*\mathcal{G}$ the structure of a right $\overline{\boldsymbol{\mathcal{A}}}$-Hopf algebra;
5. on $H_*\mathcal{E}$ the structure of an $H_*\mathcal{G}$-coalgebra and an $\overline{\boldsymbol{\mathcal{A}}}$-coalgebra;
6. on $K_*\mathcal{F}$ the structure of an $H_*\mathcal{G}$-algebra and an $\overline{\boldsymbol{\mathcal{A}}}$-algebra.

That $H_*\mathcal{G}$ is a right $\overline{\boldsymbol{\mathcal{A}}}$-Hopf algebra is given in Proposition 1.119. That $H_*\mathcal{E}$ is a $\overline{\boldsymbol{\mathcal{A}}}$-coalgebra is given in Proposition 1.118. That $H_*\mathcal{E}$ is an $H_*\mathcal{G}$-coalgebra and that $K_*\mathcal{F}$ an $H_*\mathcal{G}$-algebra follow from Propositions 1.102 and 1.31 respectively; these facts have in any case already been used in Definition 7.3. That $K_*\mathcal{F}$ is an $\overline{\boldsymbol{\mathcal{A}}}$-algebra follows from Proposition 1.31, with the roles of "left" and "right" reversed. So we can indeed apply Proposition 3.21 with the substitutions (7.3) to conclude:

PROPOSITION 7.6. *The vertical Steenrod squares on* $\mathrm{Ext}^{*,*}_{H_*\mathcal{G}}(H_*\mathcal{E}, K_*\mathcal{F})$ *satisfy the Cartan formula. They are related to the diagonal squaring operations by the equation*

$$Sq^l_V Sq^k_D = Sq^k_D Sq^{l/2}_V$$

for all $k, l \geq 0$.

3. The Operations at the E_2-level

We continue to fix a simplicial group $\mathcal{G}$, a principal right $\mathcal{G}$-space $\mathcal{E}$ that is also a Kan complex, and a Kan complex $\mathcal{F}$ on which $\mathcal{G}$ acts from the left. We continue our study of the products and Steenrod operations in the Eilenberg-Moore spectral sequence for $H^*(\mathcal{E} \times_{\mathcal{G}} \mathcal{F}; \mathbb{Z}/2)$ as they have been defined in Theorem 7.2. Our goal is to show that the product and squaring operations at the E_2-level are those that were described in purely algebraic terms in the previous section. The main results of this section are Propositions 7.7 and 7.10. We begin with the diagonal squares.

PROPOSITION 7.7. *The diagonal Steenrod operations* $Sq^k_D : E_2^{p,q} \to E_2^{p+k,2q}$ *in the Eilenberg-Moore spectral sequence agree with the diagonal operations given by Definition 7.3.*

PROOF. We have obtained the Eilenberg-Moore spectral sequence (6.1) as the spectral sequence of the bisimplicial coalgebra $X = W^G(E) \times_G F$. So in the manner of (1.149), an element $\alpha \in E_2^{p,q}$ is represented by a cochain

$$x : (W^G(E) \times_G F)_{p,q} \to \mathbb{Z}/2$$

satisfying $\delta_v x = 0$, and for which there exists a cochain

$$x' : (W^G(E) \times_G F)_{p+1,q-1} \to \mathbb{Z}/2$$

satisfying $\delta_h x = \delta_v x'$. In Chapter 6 we interpreted the E_2-term of the spectral sequence by looking at the image of such x under the composition of cochain mappings

(7.6)

$$\begin{array}{ccc}
\mathrm{Hom}(C(W_p^G(E)\times_G F),\mathbb{Z}/2) & \xrightarrow{\nabla^*(W_p^G(E),F)} & \mathrm{Hom}(CW_p^G(E)\otimes_{CG} CF,\mathbb{Z}/2) \\
 & & \Big\downarrow \phi \\
 & & \mathrm{Hom}_{CG}(CW_p^G(E),\overline{\mathrm{Hom}}(CF,\mathbb{Z}/2))
\end{array}$$

where ϕ and $\nabla^*(W_p^G(E),F)$ are as in (1.50) and (1.94) respectively. We write $\tilde{x} = \phi\,\nabla^*\,(W_p^G(E),F)(x)$ for the image of x under the composition (7.6). This cocycle lives in the cochain complex $\mathrm{Hom}_{CG}(CW_p^G(E),\overline{\mathrm{Hom}}(CF,\mathbb{Z}/2))$, and so represents an element $\{\tilde{x}\}\in H^q(\mathrm{Hom}_{CG}(CW_p^G(E),\overline{\mathrm{Hom}}(CF,\mathbb{Z}/2)))$. We denote by $\tilde{\tilde{x}}\in \mathrm{Hom}^q_{H_*\mathcal{G}}(W_p^{H_*\mathcal{G}}(H_*\mathcal{E}),K_*\mathcal{F})$ the image of $\{\tilde{x}\}$ under the mapping h of (1.49)

(7.7)

$$\begin{array}{ccc}
H^q(\mathrm{Hom}_{CG}(CW_p^G(E),\overline{\mathrm{Hom}}(CF,\mathbb{Z}/2))) & \xrightarrow{h} & \mathrm{Hom}^q_{H_*\mathcal{G}}(H_*\mathcal{W}_p^{\mathcal{G}}(\mathcal{E}),K_*\mathcal{F}) \\
 & & \Big\downarrow = \\
 & & \mathrm{Hom}^q_{H_*\mathcal{G}}(W_p^{H_*\mathcal{G}}(H_*\mathcal{E}),K_*\mathcal{F})
\end{array}$$

where we have used the isomorphism $H_*\mathcal{W}_p^{\mathcal{G}}(\mathcal{E}) = W_p^{H_*\mathcal{G}}(H_*\mathcal{E})$ as in (6.13). Finally, we denote by $\{\tilde{\tilde{x}}\}$ the class in $\mathrm{Ext}^{p,q}_{H_*\mathcal{G}}(H_*\mathcal{E},K_*\mathcal{F})$ that is represented by $\tilde{\tilde{x}}$. It was through the association of α with $\{\tilde{\tilde{x}}\}$ that we made in Chapter 6 the identification of the E_2-term of the spectral sequence with $\mathrm{Ext}^{*,*}_{H_*\mathcal{G}}(H_*\mathcal{E},K_*\mathcal{F})$. Similarly $Sq^k_D\alpha\in E_2^{p+k,2q}$ is represented by a cochain $y:(W^G(E)\times_G F)_{p+k,2q}\to\mathbb{Z}/2$, where y is given in terms of x by (2.61). We write

$$\tilde{y} = \phi\,\nabla^*\,(W^G_{p+k}(E),F)(y)\in \mathrm{Hom}^{2q}_{CG}(CW^G_{p+k}(E),\overline{\mathrm{Hom}}(CF,\mathbb{Z}/2))$$

for the image of y under the composition:

(7.8)

$$\begin{array}{ccc}
\mathrm{Hom}(C(W^G_{p+k}(E)\times_G F),\mathbb{Z}/2) & \xrightarrow{\nabla^*(W^G_{p+k}(E),F)} & \mathrm{Hom}(CW^G_{p+k}(E)\otimes_{CG} CF,\mathbb{Z}/2) \\
 & & \Big\downarrow \phi \\
 & & \mathrm{Hom}_{CG}(CW^G_{p+k}(E),\overline{\mathrm{Hom}}(CF,\mathbb{Z}/2)).
\end{array}$$

We write $\{\tilde{y}\}\in H^{2q}(\mathrm{Hom}_{CG}(CW^G_{p+k}(F),\overline{\mathrm{Hom}}(CF,\mathbb{Z}/2)))$ for the cohomology class represented by $\tilde{y}$, and we denote by $\tilde{\tilde{y}}\in\mathrm{Hom}^{2q}_{H_*\mathcal{G}}(W^{H_*\mathcal{G}}_{p+k}(H_*\mathcal{E}),K_*\mathcal{F})$ the image of $\{\tilde{y}\}$ under the mapping h, as in (7.7). Finally, we denote by $\{\tilde{\tilde{y}}\}$ the class in $\mathrm{Ext}^{p+k,2q}_{H_*\mathcal{G}}(H_*\mathcal{E},K_*\mathcal{F})$ that is represented by $\tilde{\tilde{y}}$. Our next task is to show that $\{\tilde{\tilde{y}}\} = Sq^k\{\tilde{\tilde{x}}\}$, where Sq^k is the operation on the cohomology of Hopf algebras defined by (1.126). But Proposition 2.27 tells how to express $\tilde{y}$ in terms of $\tilde{x}$. It gives $\tilde{y}$ as the composition of chain maps:

$$(7.9)\qquad CW^G_{p+k}(E) \xrightarrow{C\psi(W_{p+k})} C(W^G_{p+k}(E)\times W^G_{p+k}(E)) \xrightarrow{D_0(W^G_{p+k}(E),W^G_{p+k}(E))}$$
$$CW^G_{p+k}(E)\otimes CW^G_{p+k}(E) \xrightarrow{J^{p,p}_{p-k}(W^G(E),W^G(E))} CW^G_p(E)\otimes CW^G_p(E) -$$
$$\xrightarrow{\tilde{x}\otimes\tilde{x}} \overline{\mathrm{Hom}}(CF,\mathbb{Z}/2)\otimes\overline{\mathrm{Hom}}(CF,\mathbb{Z}/2) \xrightarrow{\mu} \overline{\mathrm{Hom}}(CF,\mathbb{Z}/2).$$

Then $\tilde{\tilde{y}} : W^{H_*\mathcal{G}}_{p+k}(H_*\mathcal{E}) \to K_*\mathcal{F}$ is the map in homology that is induced by the composition (7.9). We wish to compute this map and begin by noting that we have two descriptions of a coalgebra structure on $H_*\mathcal{W}^{\mathcal{G}}_{p+k}(\mathcal{E})$. The first is the standard description as in (1.74) and (1.76) of the coproduct on the homology of a simplicial set. This is the map induced in homology by the composition of the first two chain maps in (7.9). On the other hand we have as in Example 1.109 the simplicial object $W^{H_*\mathcal{G}}(H_*\mathcal{E})$ over the category of $H_*\mathcal{G}$-coalgebras, each of whose terms is endowed with a coproduct. In particular, $H_*\mathcal{W}^{\mathcal{G}}_{p+k}(\mathcal{E}) = W^{H_*\mathcal{G}}_{p+k}(H_*\mathcal{E})$ acquires a coproduct in this way. Clearly the coalgebra structures on $H_*\mathcal{W}^{\mathcal{G}}_{p+k}(\mathcal{E})$ we obtain by these two methods are the same: both are the tensor product of coalgebras $H_*\mathcal{E}\otimes H_*\mathcal{G}\otimes\cdots\otimes H_*\mathcal{G}$. So the map induced in homology by the composition of the first two chain maps in (7.9) can be written:

$$\psi : W^{H_*\mathcal{G}}_{p+k}(H_*\mathcal{E}) \to W^{H_*\mathcal{G}}_{p+k}(H_*\mathcal{E})\otimes W^{H_*\mathcal{G}}_{p+k}(H_*\mathcal{E}).$$

The next chain map in (7.9) is $J^{p,p}_{p-k}(W^G(E),W^G(E))$. The map it induces in homology is described by Proposition 1.126. Finally the map μ in (7.9) induces in homology the ordinary multiplication in $K_*\mathcal{F}$, as in (1.75) and (1.78). We conclude that the composition (7.9) induces in homology the composition:

$$W^{H_*\mathcal{G}}_{p+k}(H_*\mathcal{E}) \xrightarrow{\psi} W^{H_*\mathcal{G}}_{p+k}(H_*\mathcal{E})\otimes W^{H_*\mathcal{G}}_{p+k}(H_*\mathcal{E}) \xrightarrow{D^{p,p}_{p-k}(W^{H_*\mathcal{G}}(H_*\mathcal{E}),W^{H_*\mathcal{G}}(H_*\mathcal{E}))}$$
$$\to W^{H_*\mathcal{G}}_p(H_*\mathcal{E})\otimes W^{H_*\mathcal{G}}_p(H_*\mathcal{E}) \xrightarrow{\tilde{\tilde{x}}\otimes\tilde{\tilde{x}}} K_*\mathcal{F}\otimes K_*\mathcal{F} \xrightarrow{\mu} K_*\mathcal{F}.$$

So this composition is the mapping $\tilde{\tilde{y}}$. But now one can see from the definitions (1.109),(1.113), and (1.126) that $\{\tilde{\tilde{y}}\} = Sq^k\{\tilde{\tilde{x}}\}$ in $\mathrm{Ext}^{p+k,2q}_{H_*\mathcal{G}}(H_*\mathcal{E},K_*\mathcal{F})$, so we are done. □

PROPOSITION 7.8. *The product $E_2^{p,q}\otimes E_2^{r,s} \to E_2^{p+r,q+s}$ in the Eilenberg-Moore spectral sequence agrees with the product given by Definition 7.3.*

The proof is similar to, but easier than, the proof of the preceding proposition, and is left to the reader.

We turn now to the vertical squares.

If $\mathcal{W}$ is any right $\mathcal{G}$-space then the vector space $\mathrm{Hom}_{H_*\mathcal{G}}(H_*\mathcal{W},K_*\mathcal{F})$ becomes a left $\boldsymbol{\mathcal{A}}$-module, with action formally identical to the one in (7.4):

$$(7.10)\qquad (Sq^k\cdot f)(m) = \sum_{i+j=k} \overline{Sq^i} f(m\overline{Sq^j})$$

for each $f\in\mathrm{Hom}_{H_*\mathcal{G}}(H_*\mathcal{W},K_*\mathcal{F})$, $m\in H_*\mathcal{W}$, $k\geq 0$.

PROPOSITION 7.9. *Suppose $\mathcal{W}$ an extended right $\mathcal{G}$-space. Then the mapping $H^*(\mathcal{W} \times_{\mathcal{G}} \mathcal{F}) \to \mathrm{Hom}_{H_*\mathcal{G}}(H_*\mathcal{W}, K_*\mathcal{F})$ of (1.97) is an isomorphism of left $\boldsymbol{\mathcal{A}}$-modules.*

PROOF. That we have an isomorphism of vector spaces is already in Lemma 1.107. The isomorphism displayed there is a composition, and we show each mapping in the composition commutes with $\boldsymbol{\mathcal{A}}$-action. That

$$\mathrm{Hom}(H_*(\mathcal{X} \times \mathcal{F}), \mathbb{Z}/2) \to \mathrm{Hom}(H_*\mathcal{X} \otimes H_*\mathcal{F}, \mathbb{Z}/2)$$

is $\boldsymbol{\mathcal{A}}$-linear follows from Proposition 1.117. That

$$\mathrm{Hom}(H_*\mathcal{X} \otimes H_*\mathcal{F}, \mathbb{Z}/2) \to \mathrm{Hom}(H_*\mathcal{X}, \overline{\mathrm{Hom}}(H_*\mathcal{F}, \mathbb{Z}/2))$$

is $\boldsymbol{\mathcal{A}}$-linear is an application of Example 3.7 with $\Pi = \overline{\boldsymbol{\mathcal{A}}}$, $\Xi = \mathbb{Z}/2$. □

PROPOSITION 7.10. *The vertical Steenrod operations $Sq_V^k : E_2^{p,q} \to E_2^{p,q+k}$ in the Eilenberg-Moore spectral sequence agree with the vertical operations given by Definition 7.5.*

PROOF. We have obtained the Eilenberg-Moore spectral sequence (6.1) as the spectral sequence of the bisimplicial coalgebra $X = W^G(E) \times_G F$. So in the manner of (1.149), an element $\alpha \in E_2^{p,q}$ is represented by a cochain

$$x : (W^G(E) \times_G F)_{p,q} \to \mathbb{Z}/2$$

satisfying $\delta_v x = 0$, and for which there exists a cochain

$$x' : (W^G(E) \times_G F)_{p+1,q-1} \to \mathbb{Z}/2$$

satisfying $\delta_h x = \delta_v x'$. In particular we can consider that x defines an element $\tilde{x} \in H^q(\mathcal{W}_p^{\mathcal{G}}(\mathcal{E}) \times_{\mathcal{G}} \mathcal{F})$. It follows from the description of Sq_V^k given in (2.60) that $Sq_V^k\alpha$ is represented by a cochain $y : (W^G(E) \times_G F)_{p,q+k} \to \mathbb{Z}/2$, for which the associated element $\tilde{y} \in H^{q+k}(\mathcal{W}_p^{\mathcal{G}}(\mathcal{E}) \times_{\mathcal{G}} \mathcal{F})$ is $Sq^k\tilde{x}$. We rewrite these comments using the isomorphism in Proposition 7.9. We regard x as defining an element

$$\tilde{\tilde{x}} \in \mathrm{Hom}^q_{H_*\mathcal{G}}(H_*\mathcal{W}_p^{\mathcal{G}}(\mathcal{E}), K_*\mathcal{F}),$$

and y as defining an element

$$\tilde{\tilde{y}} \in \mathrm{Hom}^{q+k}_{H_*\mathcal{G}}(H_*\mathcal{W}_p^{\mathcal{G}}(\mathcal{E}), K_*\mathcal{F}).$$

Then $\tilde{\tilde{y}} = Sq^k\tilde{\tilde{x}}$, where the $\boldsymbol{\mathcal{A}}$-action on $\mathrm{Hom}_{H_*\mathcal{G}}(H_*\mathcal{W}_p^{\mathcal{G}}(\mathcal{E}), K_*\mathcal{F})$ is as described in (7.10). Finally, as in (6.13), we can make the identification $H_*\mathcal{W}_p^{\mathcal{G}}(\mathcal{E}) = W_p^{H_*\mathcal{G}}(H_*\mathcal{E})$ for each $p \geq 0$. In the present context we view this as an isomorphism not only of modules over $H_*\mathcal{G}$ but also of modules over $\overline{\boldsymbol{\mathcal{A}}}$. So $E_2^{p,*}$ appears as the p'th homology group of the complex of $\boldsymbol{\mathcal{A}}$-modules:

$$\cdots \longrightarrow \mathrm{Hom}_{H_*\mathcal{G}}(W_{p-1}^{H_*\mathcal{G}}(H_*\mathcal{E}), K_*\mathcal{F}) \xrightarrow{\delta_h} \mathrm{Hom}_{H_*\mathcal{G}}(W_p^{H_*\mathcal{G}}(H_*\mathcal{E}), K_*\mathcal{F}) \longrightarrow \cdots$$

The $\boldsymbol{\mathcal{A}}$-action on each term of the complex is given by (7.10), and the vertical squares on $E_2^{p,*}$ are given by the induced action of $\boldsymbol{\mathcal{A}}$ on the homology of the complex. But as we have described in Definition 1.45 and Example 1.110, the chain complex $CW^{H_*\mathcal{G}}(H_*\mathcal{E})$ is a resolution of the $\overline{\boldsymbol{\mathcal{A}}} - H_*\mathcal{G}$ module $H_*\mathcal{E}$ by relative projectives. We conclude that the vertical squares in the Eilenberg-Moore spectral sequence, as described in (2.60), are the same as those described in purely algebraic terms in Definition 7.5. □

4. Applications

Using the theory of Steenrod operations in the Eilenberg-Moore spectral sequence, as it was originally worked out in [**77, 78, 61, 62**], Mimura and Mori calclulated in [**54**] the cohomology rings of the exceptional Lie groups E_6 and E_7, with coefficients in $\mathbb{Z}/2$. They study the Eilenberg-Moore spectral sequences:

$$E_2^{p,q} = \mathrm{Ext}^{p,q}_{H_*E_6}(\mathbb{Z}/2, \mathbb{Z}/2) \Rightarrow H^{p+q}(BE_6; \mathbb{Z}/2) \tag{7.11}$$

$$E_2^{p,q} = \mathrm{Ext}^{p,q}_{H_*E_7}(\mathbb{Z}/2, \mathbb{Z}/2) \Rightarrow H^{p+q}(BE_7; \mathbb{Z}/2). \tag{7.12}$$

Using squaring operations in these spectral sequences they show that both collapse at the E_2-level. Only the vertical Steenrod squares are needed to prove collapse of (7.11), but the proof of collapse for (7.12) involves the diagonal operations as well. Mimura and Mori are then able to calculate the cohomology rings $H^*(BE_6; \mathbb{Z}/2)$ and $H^*(BE_7; \mathbb{Z}/2)$. These authors also obtain information about the action of the Steenrod algebra on $H^*(BE_6; \mathbb{Z}/2)$ and $H^*(BE_7; \mathbb{Z}/2)$, by using Steenrod operations in the spectral sequences. M. Mimura has informed this writer that he and Nishimoto can also prove collapse of the spectral sequence:

$$E_2^{p,q} = \mathrm{Ext}^{p,q}_{H_*E_8}(\mathbb{Z}/2, \mathbb{Z}/2) \Rightarrow H^{p+q}(BE_8; \mathbb{Z}/2) \tag{7.13}$$

by the use of Steenrod operations. Some preliminary work by these authors on $H^*(BE_8; \mathbb{Z}/2)$ can be found in [**55**].

Mimura and Sambe use Steenrod operations in the Eilenberg-Moore spectral sequence at odd primes to prove collapse of the spectral sequences for $H^*(BE_6; \mathbb{Z}/3)$ and $H^*(BE_7; \mathbb{Z}/3)$. These results are developed in the papers [**56, 57, 70, 71, 72**].

Recently Kameko and Mimura [**31**], [**32**] have analyzed the spectral sequence

$$E_2^{p,q} = \mathrm{Ext}^{p,q}_{H_*E_8}(\mathbb{Z}/3, \mathbb{Z}/3) \Rightarrow H^{p+q}(BE_8; \mathbb{Z}/3). \tag{7.14}$$

and have obtained the surprising result that it *does not* collapse. In a forthcoming paper these writers use Steenrod operations in the spectral sequence to discuss the higher differentials, and so are able to calculate $H^*(BE_8; \mathbb{Z}/3)$, [**33**].

In their Memoir [**38**], Kuribayashi, Mimura and Nishimoto study the cohomology of the spaces BLG, the classifying spaces of the loop groups of compact Lie groups G. Among their results is a computation of the cohomology rings $H^*(BL\,\mathrm{Spin}(n); \mathbb{Z}/2)$ for some low values of n. The computation uses Steenrod operations in the Eilenberg-Moore spectral sequences for these cohomology rings.

Steenrod operations in the Eilenberg-Moore spectral sequence (6.1) seem to have been applied so far only to cases in which the free $\mathcal{G}$-space $\mathcal{E}$ is contractible, and $\mathcal{F}$ is a point. It would be very interesting to see applications of the general theory developed here to cases in which $\mathcal{E}$ is not contractible.

Bibliography

[1] J.F. Adams, *On the structure and applications of the Steenrod algebra*, Comment. Math. Helv. **32** (1958), 180-214.

[2] J. Adem, *The iteration of Steenrod squares in algebraic topology*, Proc. Natl. Acad. Sci. U.S.A. **38** (1952), 720-726.

[3] J. Adem, *The relations on Steenrod powers of cohomology classes*, in: Algebraic Geometry and Topology, a Symposium in Honor of Solomon Lefscheftz, Princeton University Press, Princeton, N.J., (1957), 191-238.

[4] D. Anderson and D. Davis, *A vanishing theorem in homological algebra*, Comment. Math. Helv. **48** (1973), 318-327.

[5] M. André, *Homologie des Algèbres Commutatives*, Grundlehren der mathematischen Wissenschaften **206**, Springer Verlag, Berlin, (1974).

[6] S. Araki, *Steenrod reduced powers in the spectral sequence associated to a fibering I,II.* Mem. Fac. Sci. Kyushu Univ. Series(A) Math. **11** (1957), 15-64, 81-97.

[7] M.G. Barratt, V.K.A.M. Gugenheim, and J.C. Moore, *Semisimplicial fiber bundles*, Amer. J. Math. **81** (1959), 639-657.

[8] M. Barr and J. Beck, *Homology and standard constructions*, in: Seminar on Triples and Categorical Homology Theory, Lecture Notes in Mathematics Vol. **80**, Springer Verlag, Berlin, (1969), 245-335.

[9] D. Benson, *Modular representation theory: New trends and methods.* Lecture Notes in Mathematics **108**, Springer Verlag, Berlin, (1984).

[10] D. Benson, *Representations and cohomology II: Cohomology of groups and modules*, Cambridge studies in advanced mathematics **31**, Cambridge University Press, 1991.

[11] D. Benson and J. Carlson, *The cohomology of extra-special groups*, Bull. London Math. Soc. **24** (1992), 209-235.

[12] J. Carlson, L. Townsley, L. Valero-Elizondo, M. Zhang, *Cohomology Rings of Finite Groups.* Kluwer Academic Publishers, Dordrecht, The Netherlands, 2003.

[13] H. Cartan, *Sur l'itération des opérations de Steenrod*, Comment. Math. Helv. **29** (1955), 40-58.

[14] H. Cartan, *DGA-modules (suite); notion de construction*, Séminaire Henri Cartan 7ième année: 1954/55, Exposé 3, Ecole Normale Supérieure, Secrétariat mathématique, Paris, 1956.

[15] H. Cartan and S. Eilenberg, *Homological Algebra*, Princeton University Press, Princeton, New Jersey, 1956.

[16] A. Dold, *Über die Steenrodschen Kohomologieoperationen*, Ann. of Math.**73** (1961), 258-294.

[17] A. Dress, *Zur Spectralsequenz von Faserungen*, Invent. Math. **3** (1967), 172-178.

[18] W. Dwyer, *Higher divided squares in second-quadrant spectral sequences*, Trans. Amer. Math. Soc. **260** (1980), 437-447.

[19] S. Eilenberg and S. MacLane, *On the groups* $H(\pi, n), I$, Ann. of Math. **58** (1953), 55-106.

[20] S. Eilenberg and S. MacLane, *On the groups* $H(\pi, n), II$, Ann. of Math. **60** (1954), 49-139.

[21] S. Eilenberg and J.C. Moore, *Foundations of Relative Homological Algebra*, Mem. Amer. Math. Soc. **55**, American Mathematical Society, Providence, R.I., 1965.

[22] S. Eilenberg and J.C. Moore, *Adjoint functors and triples*, Ill. J. Math. **9** (1965), 381-398.

[23] S. Eilenberg and J. Zilber, *On products of complexes*, Amer. J. Math. **75** (1953), 200-204.

[24] M. Feth, *Erweiterungen von Bialgebren und ihre kohomologische Beschreibung*, Diplomarbeit Universitat München (1982).

[25] P. Goerss, *On the André-Quillen cohomology of commutative* F_2*-algebras*, Astérisque **186** (1990).

[26] P. Goerss, *André-Quillen cohomology and the Bousfield- Kan spectral sequence*, Astérisque **191** (1990), 109-210.

[27] V.K.A.M. Gugenheim, *On extensions of algebras, coalgebras, and Hopf algebras I* , Amer. J. of Math. **84** (1962),349-382.

[28] A. Heller, *Homological algebra in abelian categories*, Ann. of Math. **68** (1958), 484-525.

[29] G. Hochschild and J.-P. Serre, *Cohomology of group extensions*, Trans. Amer. Math. Soc. **74** (1953), 110-134.

[30] I. Hofstetter, *Extensions of Hopf algebras and their cohomological description*, J. Alg. **164** (1994), 264-298.

[31] M. Kameko and M. Mimura, *On the Rothenberg-Steenrod spectral sequence for mod 3 cohomology of the classifying space of the exceptional Lie group* E_8, RIMS Kokyuroku 1357 (February, 2004), RIMS of Kyoto University, 95-103 (in Japanese).

[32] M. Kameko and M. Mimura, *On the Rothenberg-Steenrod spectral sequence for mod 3 cohomology of the classifying space of the exceptional Lie group* E_8, to appear in "Proceedings of the Kinosaki Conference, 2003", Geometry and Topology Monographs, Mathematics Department of the University of Warwick, http://www.maths.warwick.ac.uk/gt/gtmono.html.

[33] M. Kameko and M. Mimura, to appear.

[34] L. Kristensen, *On the cohomology of two-stage Postnikov systems*, Acta Math. **107** (1962), 73-123.

[35] L. Kristensen, *On secondary cohomology operations*, Colloquium on algebraic topology, Aarhus University (1962), 16-21.

[36] L. Kristensen, *On a Cartan formula for secondary cohomology operations*, Math. Scand. **16** (1965), 97-115.

[37] T. Kudo, *A transgression theorem*, Mem. Fac. Sci. Kyushu Univ. (A) **9** (1956), 79-81.

[38] K.Kuribayashi, M.Mimura, T. Nishimoto, *Twisted tensor products related to the cohomology of the classifying spaces of loop groups*, Mem. Amer. Math. Soc. **180/849** (2006).

[39] A. Liulevicius, *The cohomology of a subalgebra of the Steenrod algebra*, Trans. Amer. Math. Soc. **104** (1962), 443-449.

[40] A. Liulevicius, *The factorization of cyclic reduced powers by secondary cohomology operations*, Mem. Amer. Math. Soc. **42** (1962), American Mathematical Society, Providence, RI.

[41] R.C. Lyndon, *The cohomology theory of group extensions*, Duke Math. J. **15** (1948), 271-292.

[42] S. Mac Lane, *Constructions simpliciales acycliques*, Colloque Henri Poincaré, Paris, 1954.

[43] S. Mac Lane, *Homology*, Springer-Verlag New York Inc., 1967.

[44] M. Mahowald and P. Shick, *Periodic phenomena in the classical Adams spectral sequence*, Trans. Amer. Math. Soc. **300** (1987), 191-206.

[45] W.S. Massey, *Products in exact couples*, Ann. of Math. **59** (1954), 558-569.

[46] W.S. Massey, *Some problems in algebraic topology and the theory of fiber bundles*, Ann. of Math. **62** (1955), 327-359.

[47] J.P. May, *Simplicial Objects in Algebraic Topology*, D. Van Nostrand Company, Princeton, New Jersey, 1967.

[48] J.P. May, *A general algebraic approach to Steenrod operations*, in: The Steenrod Algebra and its Applications, Lecture Notes in Math. **168**, Springer Verlag, Berlin-Heidelberg-New York, (1970), 153-231.

[49] H.R. Miller, *The Sullivan conjecture*, Bull. Amer. Math. Soc. **9** (1983), 75-78.

[50] H.R. Miller, *The Sullivan conjecture on maps from classifying spaces*, Ann. of Math. **120** (1984), 39-87.

[51] J.W. Milnor, J. C. Moore, *On the structure of Hopf algebras*, Ann. of Math. **81** (1965), 211-264.

[52] J.W. Milnor, *The Steenrod algebra and its dual*, Ann. of Math. **67** (1958), 150-171.

[53] M. Mimura, *Characteristic classes for the exceptional Lie groups.* In: *Adams memorial symposium on Algebraic Topology* Vol. 1, N. Ray and G. Walker, eds., Cambridge University Press 1992, 103-130.

[54] M.Mimura and M.Mori, *The squaring operations in the Eilenberg-Moore spectral sequence and the classifying space of an associative H-space, I*, Publ. RIMS Kyoto Univ. **13** (1977), 755-776.

[55] M. Mimura and T. Nishimoto, *On the cohomology mod 2 of the classifying space of the exceptional Lie group* E_8, to appear in "Proceedings of the International Conference on Algebraic

Topology, Hanoi, 2004", Geometry and Topology Monographs, Mathematics Department of the University of Warwick, http://www.maths.warwick.ac.uk/gt/gtmono.html.

[56] M. Mimura and Y. Sambe, *On the cohomology mod-p of the classifying spaces of the exceptional Lie groups, IV*, in preparation.

[57] M. Mimura, Y.Sambe, M. Tezuka, *Some remarks on the mod-3 cohomology of the classifying space of the exceptional Lie group* E_6, Proceedings of the Workshop on Pure and Applied Mathematics (Daewoo) **17**, no. 3 (1997).

[58] S. Montgomery, *Hopf algebras and their actions on rings*, Conference Board of the Mathematical Sciences, Regional Conference Series **82**, American Mathematical Society, 1993.

[59] J.C. Moore, *Seminar on algebraic homotopy theory*, (mimeographed lecture notes), Princeton University, 1956.

[60] J.C. Moore, *Algèbre homologique et homologie des espaces classifiants*, Séminaire Henri Cartan 12ième année: 1959/60, fasc.1, Exposé 7, Ecole Normale Supérieure, Secrétariat mathématique, Paris, 1961.

[61] M. Mori, *A note on Steenrod operations in the Eilenberg-Moore spectral sequence*, Proc. Jap. Acad. **53**(1977), 112-114.

[62] M. Mori, *The Steenrod operations in the Eilenberg-Moore spectral sequence*, Hiroshima Math. J. **9** (1979), 17-34.

[63] B. Pachuashvili, *Cohomologies and extensions in monoidal categories*, J. Pure and App. Alg. **72** (1991), 109-147.

[64] J. Palmieri, *Quillen stratification for the Steenrod algebra*, Ann. of Math. **149** (1999), 421-449.

[65] D. Quillen, *Spectral sequences of a double semi-simplicial group*, Topology **5** (1966), 155-157.

[66] D. Quillen, *On the cohomology of commutative rings*, Proc. Symp. Pure Math. **17**, Amer. Math. Soc., Providence, RI (1970), 65-87.

[67] D. Quillen, *The spectrum of an equivariant cohomology ring I,II*, Ann. of Math. **94** (1971), 549-572 and 573-602.

[68] D. Quillen, *The mod-2 cohomology of extra-special 2-groups and the spinor groups*, Math. Ann. **194** (1971), 197-212.

[69] D. Rector, *Steenrod operations in the Eilenberg-Moore spectral sequence*, Comment. Math. Helv. **45** (1970), 540-552.

[70] Y. Sambe, *A note on the cobar construction and the twisted tensor product*, Bull. Asso. of Nat. Sci., Senshu Univ. **22** (1991), 1-12.

[71] Y. Sambe, *A routine procedure to obtain elements in the cobar construction I.*, Bull. Asso. of Nat. Sci., Senshu Univ. **22** (1991), 13-18.

[72] Y. Sambe, *A routine procedure to obtain elements in the cobar construction II., III., IV.*, Bull. Asso. of Nat. Sci., Senshu Univ. **23** (1992).

[73] J. Sawka, *Odd primary Steenrod operations in first quadrant spectral sequences*, Trans. Amer. Math. Soc. **273** (1982), 737-752.

[74] J.-P. Serre, *Homologie singulière des espaces fibrés*, Ann. of Math. **54** (1951), 425-505.

[75] N. Shimada and A. Iwai, *On the cohomology of some Hopf algebras*, Nagoya Math. J. **30** (1967), 103-111.

[76] W.M. Singer, *Extension theory for connected Hopf algebras*, J. Alg. **21**, 1972, 1-16.

[77] W.M. Singer, *Steenrod squares in spectral sequences I*, Trans. Amer. Math. Soc. **175** (1973), 327-336.

[78] W.M. Singer, *Steenrod squares in spectral sequences II*, Trans. Amer. Math. Soc. **175** (1973), 337-353.

[79] W.M. Singer, *On the algebra of operations for Hopf cohomology*, Bull. London Math. Soc. **37** (2005), 627-635.

[80] L. Smith, *Homological algebra and the Eilenberg-Moore spectral sequence*, Trans. Amer. Math. Soc. **129** (1967), 58-93.

[81] L. Smith, *On the Künneth theorem I*, Math. Zeit. **116** (1970), 94-140.

[82] E.H. Spanier, *Algebraic Topology*, McGraw-Hill, 1966.

[83] N.E. Steenrod, *Products of cocycles and extensions of mappings*, Ann. of Math. **48** (1947), 290-320.

[84] N.E. Steenrod, *The Topology of Fiber Bundles*, Princeton University Press, Princeton, NJ, 1951.

[85] N. Steenrod and D.B.A. Epstein, *Cohomology Operations*, Princeton University Press, Princeton, N.J., 1962.

[86] H. Uehara, *Algebraic Steenrod operations in the spectral sequence associated with a pair of Hopf algebras*, Os972), 131-141.
[87] R. Vazquez-Garcia, *Note on Steenrod squares in the spectral sequence of a fiber space*, Bol. Soc. Math. Mex. (2) **2** (1957) 1-8 (in Spanish).
[88] C. Wilkerson, *The cohomology algebras of finite dimensional Hopf algebras*, Trans. Amer. Math. Soc **264** (1981), 137-150.

Index

Titles in This Series

129 **William M. Singer,** Steenrod squares in spectral sequences, 2006

128 **Athanassios S. Fokas, Alexander R. Its, Andrei A. Kapaev, and Victor Yu. Novokshenov,** Painlevé transcendents, 2006

127 **Nikolai Chernov and Roberto Markarian,** Chaotic billiards, 2006

126 **Sen-Zhong Huang,** Gradient inequalities, 2006

125 **Joseph A. Cima, Alec L. Matheson, and William T. Ross,** The Cauchy Transform, 2006

124 **Ido Efrat, Editor,** Valuations, orderings, and Milnor K-Theory, 2006

123 **Barbara Fantechi, Lothar Göttsche, Luc Illusie, Steven L. Kleiman, Nitin Nitsure, and Angelo Vistoli,** Fundamental algebraic geometry: Grothendieck's FGA explained, 2005

122 **Antonio Giambruno and Mikhail Zaicev, Editors,** Polynomial identities and asymptotic methods, 2005

121 **Anton Zettl,** Sturm-Liouville theory, 2005

120 **Barry Simon,** Trace ideals and their applications, 2005

119 **Tian Ma and Shouhong Wang,** Geometric theory of incompressible flows with applications to fluid dynamics, 2005

118 **Alexandru Buium,** Arithmetic differential equations, 2005

117 **Volodymyr Nekrashevych,** Self-similar groups, 2005

116 **Alexander Koldobsky,** Fourier analysis in convex geometry, 2005

115 **Carlos Julio Moreno,** Advanced analytic number theory: L-functions, 2005

114 **Gregory F. Lawler,** Conformally invariant processes in the plane, 2005

113 **William G. Dwyer, Philip S. Hirschhorn, Daniel M. Kan, and Jeffrey H. Smith,** Homotopy limit functors on model categories and homotopical categories, 2004

112 **Michael Aschbacher and Stephen D. Smith,** The classification of quasithin groups II. Main theorems: The classification of simple QTKE-groups, 2004

111 **Michael Aschbacher and Stephen D. Smith,** The classification of quasithin groups I. Structure of strongly quasithin K-groups, 2004

110 **Bennett Chow and Dan Knopf,** The Ricci flow: An introduction, 2004

109 **Goro Shimura,** Arithmetic and analytic theories of quadratic forms and Clifford groups, 2004

108 **Michael Farber,** Topology of closed one-forms, 2004

107 **Jens Carsten Jantzen,** Representations of algebraic groups, 2003

106 **Hiroyuki Yoshida,** Absolute CM-periods, 2003

105 **Charalambos D. Aliprantis and Owen Burkinshaw,** Locally solid Riesz spaces with applications to economics, second edition, 2003

104 **Graham Everest, Alf van der Poorten, Igor Shparlinski, and Thomas Ward,** Recurrence sequences, 2003

103 **Octav Cornea, Gregory Lupton, John Oprea, and Daniel Tanré,** Lusternik-Schnirelmann category, 2003

102 **Linda Rass and John Radcliffe,** Spatial deterministic epidemics, 2003

101 **Eli Glasner,** Ergodic theory via joinings, 2003

100 **Peter Duren and Alexander Schuster,** Bergman spaces, 2004

99 **Philip S. Hirschhorn,** Model categories and their localizations, 2003